Elastic and Plastic Deformation of Carbon Nanotubes

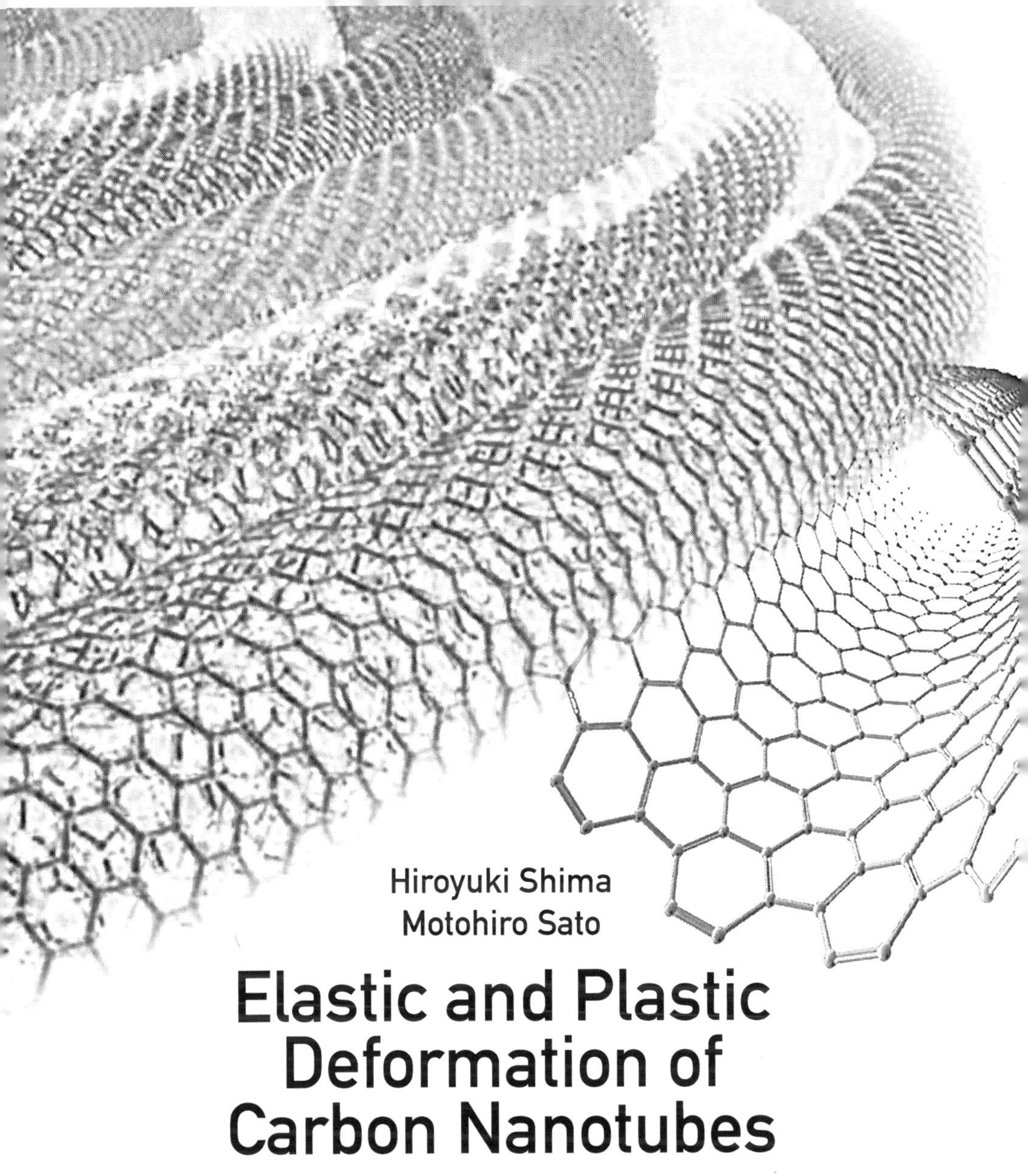

Hiroyuki Shima
Motohiro Sato

Elastic and Plastic Deformation of Carbon Nanotubes

Pan Stanford Publishing

Published by

Pan Stanford Publishing Pte. Ltd.
Penthouse Level, Suntec Tower 3
8 Temasek Boulevard
Singapore 038988

Email: editorial@panstanford.com
Web: www.panstanford.com

British Library Cataloguing-in-Publication Data
A catalogue record for this book is available from the British Library.

Elastic and Plastic Deformation of Carbon Nanotubes

ISBN 978-981-4310-75-8 (Hardcover)
ISBN 978-981-4364-15-7 (eBook)

Printed in the USA

To Rena and Shiho

Contents

Preface

The key players in the materials field continually change with the times. The 19th century saw the rise of iron, followed by silicon in the 20th century. In the 21st century, nano-carbon materials look set to take center-stage.

Carbon is a rare substance that can take various structures and forms. When carbon atoms form a three-dimensional structure, their glittering beauty as diamonds captivates people. When aligned in a two-dimensional plane, carbon atoms become just black graphite and lose their sparkle. In addition to these two macroscale carbon materials, several other nano-carbon materials have been discovered in just the last 20 years, opening new horizons in physics and chemistry. It all began with the C_{60} molecule (fullerene), whose existence was predicted by Osawa in 1970 and evidenced by Kroto *et al.* in 1985. Subsequent studies, including those on carbon nanotubes by Iijima in 1991 and on graphene by Novoselov *et al.* in 2004, have had a tremendous impact, driving developments in science and engineering around the turn of the century.

Of these three types of nano-carbon materials, carbon nanotubes are attracting greatest attention in both industry and academia. Research on carbon nanotubes has focused on two characteristics not usually seen in other fields. First and foremost is the sheer breadth of the research, which encompasses physics, chemistry, materials science, electric and electronic engineering, and life science. The second characteristic is that basic research and applied research are extremely close to each other. A succession of the phenomena of interest to scientists has been discovered like a treasure chest, each leading to an innovative application or development. Thanks to these two characteristics, applied research extends as far as electronic devices, fuel cells, panel display materials and gas absorption and has seen dramatic progress. It is difficult even for experts in the field to understand the progress being made outside of their specialties.

One of the reasons why carbon nanotubes offer huge potential in so many fields is their mechanical properties, specifically the following four:

- **Rigidity**: Young's modulus in the longitudinal direction exceeds that of diamond.
- **Resilience**: The original structure is recovered with few defects after large bending or deformation.
- **Toughness**: Cracks do not easily form or advance, so carbon nanotubes retain their cylindrical shape without breaking even when a large external force is applied.
- **Anisotropy**: Reflecting the cylindrical carbon shell structure with concentric axis, they show mechanical properties with extremely large anisotropy.

In addition to these significant mechanical properties, studies have shown that mechanical deformation causes considerable changes in electronic, optical, magnetic and chemical properties. Thus, many studies on new technologies to utilize the correlation between deformation and properties are under way in various fields, including electronics, biotechnology, and material design. For example, nanoscale devices that use the changes in electrical conductivity or optical response due to deformation are one of the most popular fields in nanotechnology. Meanwhile, ultrahigh-strength materials are being developed for the next-generation infrastructure such as superhigh-rise buildings and large aerospace equipment by utilizing the high thermal and environmental resistance of nanotubes. The application of these low-density substances containing only light carbon atoms for aircraft and automobile parts is expected to raise fuel efficiency and save energy, as well as dramatically reduce exhaust gas emissions and environmental impact.

Despite so many new phenomena and technologies based on the mechanical functions of nanotubes in diverse fields, there are surprisingly few books and overall reviews of nanotubes. Independent books and review papers have tended to focus on specific topics such as quantum devices utilizing the mechanical properties of nanotubes and composite strength materials utilizing the mechanical strength of nanotubes. There have been few reports on the most basic characteristic of nanotubes, which is the unique applicability of their mechanical properties in so many fields.

This book draws on various documents to give a comprehensive overview of the research progress to date, focusing on the mechanical properties of carbon nanotubes. The first half examines minute deformations in the elastic region, reviewing the huge body of literature immediately after nanotubes were discovered to the latest results still in press, without assuming detailed background knowledge. The latter half takes a closer look at the major themes of common interest to researchers in diverse fields, such as the plastic deformation of nanotubes under extreme conditions, including giant deformation, high temperature and pressure, and electron beam irradiation. In introducing prior studies, we have minimized the number of citations to avoid exhaustive descriptions, and instead have included a good balance of visually appealing figures and photographs to assist the readers' understanding. To avoid redundancy with existing books, we have cited the latest studies being reported while this book was being written. We hope it will give readers a broad view of the mechanical properties of nanotubes and help them find new research themes.

The authors express cordial gratitude to Profs. Kousuke Yakubo, Takashi Mikami, Shunji Kanie, and Minoo H. Patel. Thanks to their strict yet devoted instruction in our young researcher ages, we were able to gain the skill required to complete this book. One of the authors, Shima, wrote part of this book within his stay at Universitat Politècnica de Catalunya in Barcelona, Spain, for almost a year, during which time Prof. Marino Arroyo and Dr. Susanta Ghosh were very kind to him. Shima also deeply thanks Prof. Emeritus Eiji Osawa, Profs. Jun Onoe, Hideo Yoshioka, Yoshitaka Umeno, and Dr. Shota Ono, as communications with them have sparked his interest in the study of nano-carbon materials. The other author, Sato, was able to complete this book with the help of Dr. M. Ahmer Wadee while working for half a year at Imperial College London. Jenny Rompas, Stanford Chong, and Arvind Kanswal of Pan Stanford Publishing encouraged us to write this book and provided continuous support.

Finally, we thank our beloved wives, Rena and Shiho, and our children, Yuuka, Ayaka, Tatsuya, and Kana, for easing our fatigue through their smiles and kind words every day. We are grateful to all of them for looking after us, and for the friendships between us.

Hiroyuki Shima
Kofu, Japan
Motohiro Sato
Sapporo, Japan
July 2013

Chapter 1

Introduction

He that would know what shall be must consider what has been.
— Proverb

1.1 Carbon Nanotube: The World's Stiffest Material

Can you imagine a material that is many times stronger than steel, yet flexible enough not to break when bent or twisted, that withstands chemicals and high heat, and that conducts electricity better than silver and conducts heat better than diamond. Does this dream material really exist? Yes. Carbon nanotubes.

Carbon nanotubes are cylindrical substances comprising only carbon atoms. Their hollow centers are a vacuum, and they have fine structures of 1–10 nm in diameter[a] and 1 μm to 1 mm or more in length. As their diameters are smaller than the wavelength of visible light (∼400–800 μm), they cannot be observed directly by a light

[a] 1 nm is the thickness of DNA.

Elastic and Plastic Deformation of Carbon Nanotubes
Hiroyuki Shima and Motohiro Sato

ISBN 978-981-4310-75-8 (Hardcover), 978-981-4364-15-7 (eBook)
www.panstanford.com

microscope and so must be observed under larger, more expensive electron microscopes.[a]

The amazing robustness of nanotubes is solely due to chemical bonding between carbon atoms. The cylindrical sheet (graphite surface) which forms the nanotube has a seamless structure comprising a hexagonal grid of carbon atoms. The bonds between carbon atoms (scientifically known as "sp^2 carbons") that form hexagonal shapes such as benzene rings are the strongest of all atomic bonds. A carbon nanotube consists entirely of this strongest bond, and therefore, it is extremely resistant to bending or pulling, is very stable, and reacts with few chemicals.

The excellent mechanical properties of nanotubes include an extremely high Young's modulus[b] and tensile strength.[c] Based on previous studies, Young's modulus of nanotubes is in the order of 1 TPa (see Chapter 2), and its tensile strength is 50 GPa or higher (Chapter 7). Since both these values are several times higher than those of steel and its specific gravity is less than one-tenth that of steel, it offers 100 times larger mechanical strength than steel with the same weight. In fact, nanotubes are so robust that by weaving them into a fiber, Theoretically, it would be possible to hang a weight of 1200 tons from a rope just 1 cm in diameter. Thus, it is far stronger than any existing materials.

Another characteristic is its extreme flexibility to bending and twisting deformation. Electron microscope observations have shown that it returns to its original shape like a rubber hose even

---

[a] The electron microscope is a type of microscope in which an object is observed by irradiating it with an electron beam and magnifying the transmitted or reflected electrons. Since the degree of electron beam transmission or reflection varies depending on the structure or substances of the object, the density of the transmitted or reflected electrons varies by the position and thus produces a microscopic image. It is divided broadly into two major types: transmission electron microscope (TEM) and scanning electron microscope (SEM), both of which are frequently used in carbon nanotube measurements.

[b] Young's modulus is a constant that specifies how much stress is necessary per unit distortion. For example, Young's modulus for copper is approximately 100 GPa, which causes 0.1% distortion when a 10 kg weight is hung on a copper wire with a cross section of 1 mm^2 and length of 1 m.

[c] Tensile strength is the maximum tensile stress that a material can withstand without breaking.

when the entire tube is twisted or bent until it is warped (see Chapter 5). In addition, it has far superior corrosion resistance to metals, as it is chemically very stable and does not react with acids or alkalis. It is also more resistant to temperature than other materials, retaining sufficient strength up to 1200°C provided no oxygen is present.

However, even a material as innovative as carbon nanotube has some weaknesses. The first weakness and the greatest hindrance to practical applications is the cost of production. When purchased from a manufacturer, nanotube samples with high purity (few defects) cost several hundred dollars per gram. Since even pure gold costs only about $10 per gram, this is a steep price. The second weakness is the extreme difficulty of manipulating it mechanically and identifying its structure due to the minute nanoscale structures. Since the experimental procedures are so difficult, many researchers have tried to develop theoretical and numerical calculation methods for studying the mechanical properties of nanotubes. Many of these methods can be broadly classified as either a continuum mechanics approximation (see Chapter 4) or an atomistic approach (Chapter 3). The former tries to elicit the mechanical properties by mapping nanotubes with cylindrical graphite structures onto a continuous elastic shell and utilizing the accumulated study results of macro-scale structural mechanics. The latter identifies the mechanical behavior of nanotubes through large-scale simulations including as many atoms as possible by numerically reproducing the chemical bonding potential acting between carbon–carbon atoms. While these completely different approaches might seem to be incompatible with each other, a theoretical framework to connect these approaches without contradiction has already been proposed and appears to function well.

Before proceeding to the main theme of the mechanical properties of nanotubes, let us take a look at the history of who discovered nanotubes and when. The answer seems obvious at first glance, but is actually quite profound. Since this may be the first time for younger readers to learn the facts, the next section describes the history of the discovery of nanotubes.

1.2 Discovery before 1991: Who Found It First?

1.2.1 *Matter in Issue*

As is well known, most academic and popular literature attributes the discovery of carbon nanotubes to Sumio Iijima of NEC in 1991. With no doubt, it is Iijima's seminal article in 1991 [1], entitled "Helical microtubules of graphitic carbon," that triggered the explosion of the carbon nanotube research still fascinating us today. Iijima studied the arc evaporation process that efficiently produced fullerene (C_{60}) molecules.[a] When analyzing a by-product in the arc evaporation, Iijima found large amounts of multiwalled carbon nanotubes (MWNTs) mixed with faceted graphitic particles.[b]

From a historical perspective, however, Iijima's finding is most certainly not the first that reported about carbon nanotubes. Careful analysis of the literature shows that there had been many precedents prior to 1991, which uncovered the presence of analogous (and almost identical) nanostructures including "carbon tubes" [5] and "hollow carbon fibers" [6]. The relevant literature has been probed with great care by Boehm in 1997 [7] and Monthioux and Kuznetsov in 2006 [8], and they concluded that the origin of carbon nanotubes has been misstated repeatedly in a routine manner, i.e., Iijima's work should be considered as a "re"-discovery.

The first discovery of carbon nanotube remains a subtle and contentious issue; there are as many opinions as there are people. Motivated by the background, we believe that it is a merit for readers (in particular for young colleagues) to have an opportunity of surveying the achievements of our predecessors. Although the issue is, of course, nothing to do with the carbon nanotube mechanics itself, the historical knowledge should be fundamental toward unlocking the future of carbon nanotube studies at the next-generation stage.

---

[a] A fullerene, or sometimes called a buckyball, is any molecule composed entirely of carbon, in the form of a hollow sphere or ellipsoid. The existence of fullerene was predicted by Eiji Osawa in 1970, reported in a Japanese scientific journal [2], and subsequently evidenced in experiments in 1985 by Harold Kroto, Robert Curl, Richard Smalley, and their colleague [3].

[b] The microscope images and electron diffraction patterns obtained by Iijima [1] can be viewed, for example, in a recent review [4] as reproductions.

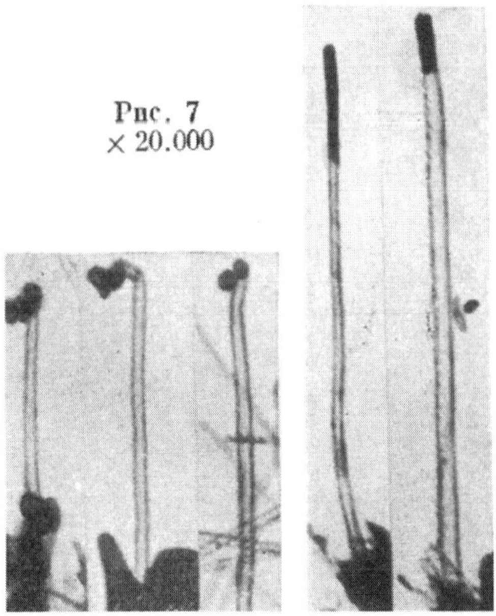

Figure 1.1 The earliest TEM images of carbon nanotubes published in Ref. [5].

1.2.2 *Historical Facts*

The first evidence for nano-sized carbon tubes is believed to have been published in 1952 in a Soviet academic Journal [5], almost 40 years (!) before Iijima's paper. The article, written by Radushkevich and Lukyanovich, demonstrated clear images of tubes (diameter ~50 nm) made of carbon using transmission electron microscopy (TEM). Figure 1.1 is a reprint of the image given in Ref. [5], which clearly shows carbon filaments exhibiting a continuous inner cavity.[a] The fact may encourage us to say that Radushkevich and Lukyanovich should be credited for the discovery of carbon nanotubes. Regrettably, this discovery was largely unnoticed, partly because the article was published in the Russian language, and

[a]Despite the scale bar omitted, the magnification value indicated in the article allows one to estimate that the diameters of the carbon tubes imaged are in the range of 50 nm, i.e., definitely nano-sized.

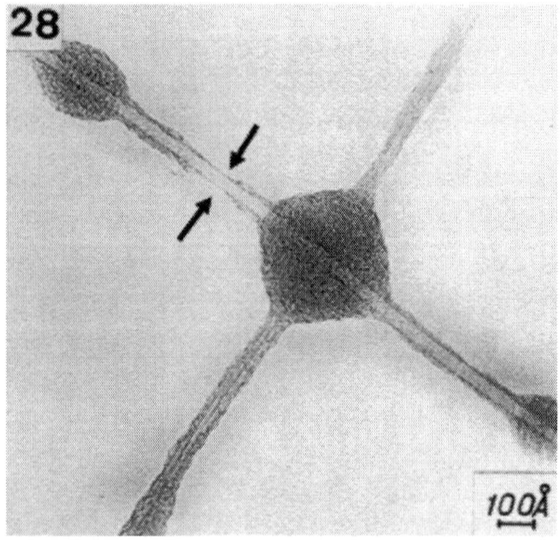

Figure 1.2 TEM image of nanotubes observed in branched inhomogeneous carbon filaments (indicated by arrows). This image is a reprint from Monthioux and Kuznetsov's review [8], but a similar one was published in the original paper of Ref. [6].

because at that time Western scientists' access to Soviet press was restricted due to the Cold War.

Subsequently, many other reports followed the Soviet work. For instance, Oberlin, Endo, and Koyama fabricated in 1976 hollow carbon fibers with nanometer-scale diameters (see Fig. 1.2) using a vapor-growth technique [6]. In 1979, Abrahamson presented evidence of carbon nanotubes at a conference paper together with the their characterization and hypotheses for the growth mechanism [9]. In 1981, a group of Soviet scientists produced carbon tubular crystals and identified them with graphene layers rolled-up into cylinders. Besides, they speculated that by rolling graphene layers into a cylinder, many different arrangements of graphene hexagonal nets are possible such as an armchair-type and a chiral one. In 1987, Tennett obtained a U.S. patent for the production of "cylindrical discrete carbon fibrils" with a "constant diameter between about 3.5 and about 70 nanometers. . . , length 102 times the diameter, and an outer region of multiple essentially continuous layers of ordered carbon atoms and a distinct inner core. . ." [10].

It is thus fair to say that carbon nanotubes were shown to exist prior to Iijima's seminal work in 1991, but no fabrication process was known that would lead to the synthesis of macroscopic amounts of carbon nanotubes. As a result, none of the earlier findings reached the impact of Iijima's 1991 paper. In addition to the loss of mass production method, Monthioux and Kuznetsov [8] pointed out that the carbon filaments had been rated just unwanted soot. Before Iijima's finding, the filaments used to be investigated by material scientists who aimed to understand the growth mechanisms so as to prevent their formation in coal and steel industry processing and in the coolant channels of nuclear reactors. On these background, the undoubted tremendous impact of the 1991 Iijima paper came from a boost received from its relation to the earlier worldwide research hit (i.e., fullerenes), and a fully mature scientific audience ready to ride a wave of nanoscience and nanotechnology.

1.2.3 *Closing Remark*

In summary, the credit for discovering carbon nanotubes should go to Radushkevich and Lukyanovich. However, merely looking with a microscope is completely different from identifying its structure. That is, Iijima not only discovered and observed the nanotube with an electron microscope but also accurately elucidated its structure from electron diffraction images, and his work was thus on a different level to preceding studies.

Let us return to the first electron micrographs of carbon nanotubes captured by Iijima [1, 4]. In those micrographs, surprisingly, the multiwalled structure with an interlayer spacing of less than 1 nm is clearly observed. Even more surprising, he measured the electron diffraction shape from a single carbon nanotube. The diffraction strength from a crystal of a light element (carbon) with a thickness of only several nanometers is naturally extremely low, and so this experiment is not easy even when using the best electron microscopes available today. Iijima must have had first-class microscope operating skills to make these measurements in 1991, although he did it with ease. It is also astounding that Iijima identified this substance, which only appeared as a thin, needle-like crystal, as a "tube-shaped substance with a helical structure" based on its electron diffraction shape. How many people in the world would have been able to conclude that it is a nanotube with a helical structure by looking at the electron diffraction shape? Now that we already know the structure of carbon nanotubes, it is not so difficult to read the electron diffraction shape. But at that time, not just coincidence but also much experience and sharp insight must have been required by the person seeing this electron diffraction shape for the first time in the world to interpret it correctly as a tube-shaped substance with a helical structure.

So, who actually discovered carbon nanotubes? To answer this question, first we need to define the meaning of "discovery of nanotubes." It is probably correct to state that Radushkevich and Lukyanovich were the first to take photographs of the overall picture of the cylindrical tube-shaped nano-carbon substance (although photographs taken by other researchers could still be discovered in future). On the other hand, Iijima was the first to reveal that the microscopic structure unique to nanotubes was multiwalled and helical, by using "the eyes" of electron diffraction. The rest is up to the philosophy and preference of the person who is judging. And whatever the judgment, we should never forget that the knowledge and benefits that we now enjoy in nanotube research are the result of the hard work and achievements of our predecessors.

Chapter 2

Young's Modulus Measurement

You must be faced with phenomenal wonders with your own eyes.
— Torahiko Terada, in "Zuihitu-syu" (in his collected essays)

2.1 Introduction

Excellent mechanical properties of carbon nanotubes originate largely from the strong sp^2-hybridized chemical bonds between carbon atoms. A hallmark of the excellence is remarkably high Young's modulus. In fact, it reaches values of about 1 TPa [11, 14], which makes carbon nanotubes one of the stiffest materials available.

Experimental methods measuring Young's modulus of carbon nanotubes are conceptually equal to those applied to macroscopic materials. In practice, however, experimental uncertainties peculiar to nanometer-scale measurements hinder a straightforward manipulation of the system considered. This chapter outlines the history of experimental efforts to overcome such difficulties in nanoscale measurements. The successful determinations of the

Elastic and Plastic Deformation of Carbon Nanotubes
Hiroyuki Shima and Motohiro Sato

ISBN 978-981-4310-75-8 (Hardcover), 978-981-4364-15-7 (eBook)
www.panstanford.com

highest stiffness inherent to carbon nanotubes have encouraged the prosperity of the nanotube research, which is still ongoing.

2.2 Thermally Induced Vibration

The first measurement of Young's modulus was performed by observing vibration amplitude of a free-standing MWNT under thermal excitation [16]. The MWNTs were embedded in amorphous material at one end, having the other end standing in space. The measurement was based on a transmission electron microscopy (TEM); the TEM was equipped with a variable temperature sample holder in order to monitor the amplitude of the thermally induced oscillations of the free-standing tip as a function of the ambient temperature

Young's modulus of the free-standing MWNTs mentioned above was evaluated by assuming that they are equivalent to hollow elastic cylinders of length L, with outer and inner radii a and b, respectively. The continuum elastic theory tells us that the horizontal vibration amplitude u_n of an n-th mode at the tip is related to Young's modulus Y and the vibration energy W_n. When limited to small vibration amplitudes, W_n reads

$$W_n = \frac{1}{2} c_n u_n^2. \tag{2.1}$$

The effective force constant (or spring constant) c_n is defined by [15]

$$c_n = \frac{\pi}{16} \cdot \frac{\beta_n^2 (a^4 - b^4)}{L^3} Y, \tag{2.2}$$

where β_n is a solution to

$$\cos \beta_n \cosh \beta_n = -1. \tag{2.3}$$

For instance, $\beta_0 \simeq 1.875$ for the fundamental mode of $n = 0$. From the law of equipartition, it follows that the average of W_n, denoted by $\langle W_n \rangle$, at room temperature is given by

$$\langle W_n \rangle = k_B T \tag{2.4}$$

for each lateral vibration mode. From Eqs. (2.1)–(2.4), therefore, Young's modulus Y is determined as the slope of the linear plot of

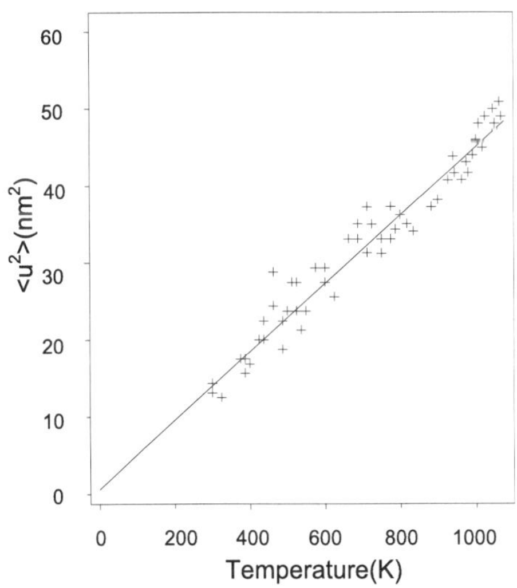

Figure 2.1 The mean-square vibration amplitude versus temperature for a 5.1 μm-long, 16.6 nm-wide, free-standing MWNT. An effective Young's modulus of 3.7 ± 0.2 TPa is deduced from the slope. Adapted from Ref. [16].

$\langle u_{\rm n}^2 \rangle$ versus T (see Fig. 2.1), i.e.,

$$\langle u_{\rm n}^2 \rangle \propto Y^{-1} T. \tag{2.5}$$

In the actual measurement reported in Ref. [16], an average value of 1.8 TPa out of 11 different nanotubes is obtained, in which individual values ranged from 0.4 to 4.15 TPa. The scatter of the data is relatively large, reflecting the experimental difficulties in obtaining accurate values of the oscillation amplitude, the inner and outer radii of the tubes, their length, etc. Nevertheless, these results provided the first proof that carbon nanotubes possess exceptionally high stiffness.

Shortly afterward, the same experimental approach was employed to measure Young's modulus of isolated single-walled carbon nanotubes (SWNTs) [17]. In this case, a larger number of samples of nanotubes (27 in total) were used, and the mean value obtained was 1.25 TPa. This value is very similar to that of the in-plane elastic constant of graphene layers.[a]

[a]Likewise, Young's modulus of 1.22 TPa was obtained for a boron nitride (BN) nanotube using the same approach [18]. This was the first measurement of Young's

2.3 Electrically Induced Vibration

Another method based on exciting the mechanical resonance of nanotubes inside a TEM has been demonstrated by Ref. [11]. In the measurement, MWNTs were excited using an alternating electric field, which induces an electrical potential difference between the nanotube and a counterelectrode.[b] See Fig. 2.2 for the experimental setup [19], which is nearly same (not identical) as that employed in Ref. [11]. The charge induced on the nanotube interacts with the electric fields between the nanotube and the counterelectrode, resulting in an attractive force that causes the nanotube to bend.[c] Hence, alternating field application leads to the nanotube vibration whose amplitude becomes maximum when the field frequency is resonant to the vibration mode of nanotubes.

Once the resonance frequencies are evaluated, Young's modulus of nanotubes Y can be calculated from the two lowest resonant frequencies: $\nu_{\rm n}$ with $n = 1, 2$ ($\nu_1 < \nu_2$). Using a formula from the Bernoulli–Euler analysis of cantilevered elastic beams [20], one obtains[d]

$$\nu_{\rm n} = \frac{\beta_{\rm n}^2}{4\pi L^2}\sqrt{\frac{Y(a^2+b^2)}{\rho}}, \tag{2.6}$$

where a and b are the outer and inner radii of nanotubes, L is the nanotube length, and ρ is the density of mass; $\beta_{\rm n}$ is a constant determined by Eq. (2.3), and $\beta_0 = 1.875$, $\beta_1 = 4.694$.

The bottom right panel in Fig. 2.2 gives a reproduction data of Young's modulus Y [11] for several MWNTs determined from

modulus of non-carbon nanotubes, and these results indicated that Young's modulus of BN nanotubes was also very high, and even competitive with that of carbon nanotubes.

[b] In most techniques discussed in this section, "electrically" induced mechanical oscillations in carbon nanotubes are detected using "non-electrical" methods. Alternatively, there is a fully electrical method that simultaneously actuates and detects mechanical oscillation in nanotubes. It is called the harmonic detection of resonance method; readers can refer to Refs. [12, 13] for details.

[c] The bending process was reversible even for extreme bends (with curvature radii < 100 nm), which indicates the large elastic strength of the nanotubes.

[d] Strictly speaking, Y in Eq. (2.6) should be replaced with the bending modulus $Y_{\rm bend}$ that may or may not equal to Y. Nevertheless, if the beam is bent by both the elongation of the outer arc and the compression of the inner arc of the bend, then $Y_{\rm bend}$ can be identified with Y of the material [21].

measurements of ν_2 and $2a$.[a] Inset shows a Lorentzian fitting for a resonance mode; the peak position is dependent on the nanotube radius and/or bending modulus, while the shape of Lorentzian curve is almost invariant to the change in those parameters. It is clear from the figure that Y is very large ($\sim$1 TPa) for $a < 10$ nm, and it drops dramatically to lower values ($Y \sim 100$ GPa) for tubes of larger diameter. Such a great reduction in Y must be related to the emergence of wavelike distortion (or called "ripple") on the inner arc of the bent nanotube. The rippling mode is known to be the energetically favorable bending mode for slightly bent, relatively thick nanotubes [22, 23], as will be explained in Section 5.4.1.

2.4 Singly Clamped Deflection

The preceding two sections have described the measurements techniques based on TEM. A different means to probe Young's modulus of carbon nanotubes was pioneered through an atomic force microscope (AFM) technique [24]. An AFM allows one to laterally bend a carbon nanotube and simultaneously record the restoring force exerted on the AFM tip as a function of the nanotubes deflection. The deformations inflicted on the nanotubes are much larger than the amplitude of the induced oscillations. Therefore, it was crucially important to fix the carbon nanotubes by one end on a substrate before they could be bent. This was done by dispersing the nanotubes on a cleaved MoS_2 substrate, and then depositing pads of SiO_x. Single-crystal MoS_2 was used as the substrate because it has a low coefficient of friction and may be cleaved to produce large, atomically flat surfaces [25].

The experimental setup and the measurement procedure are illustrated in the left panel of Fig. 2.3. The tip (blue triangle) moves in the direction of the arrow, and the lateral force is indicated by the red trace at the bottom. In the left panel, before the tip contacts the beam, the lateral force remains constant and equal to the friction force. The middle panel shows the tip bending the beam with a

[a]The inner diameter, $2b$, was not measured for all nanotubes since Y was rather insensitive to b in the conditions used.

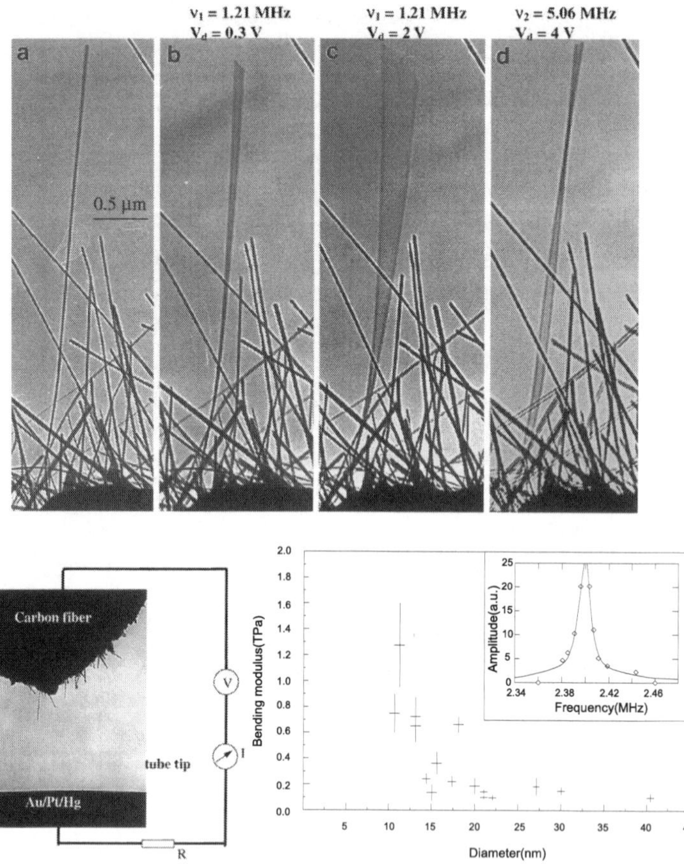

Figure 2.2 Upper panel: Field-induced vibration of a selected carbon nanotube at (a) stationary, (b,c) the first harmonic resonance at ν_1 and (d) the second harmonic resonance at ν_2. V_d is the magnitude of the applied voltage. Reproduction of Wang, Z. L. *et al. J. Phys. Chem. Sol.* **61**, 1025–1030 (2000). Bottom left: TEM image showing carbon nanotubes at the end of the electrode and the other count electrode. A constant or alternating voltage can be applied to the two electrodes to induce electrostatic deflection or mechanical resonance. Reproduction of Wang, Z. L. *et al. J. Phys. Chem. Sol.* **61**, 1025–1030 (2000). Bottom right: Young's modulus Y as a function of the outer tube diameter. Inset: Lorentzian fitting for a resonance mode. Adapted from Ref. [11].

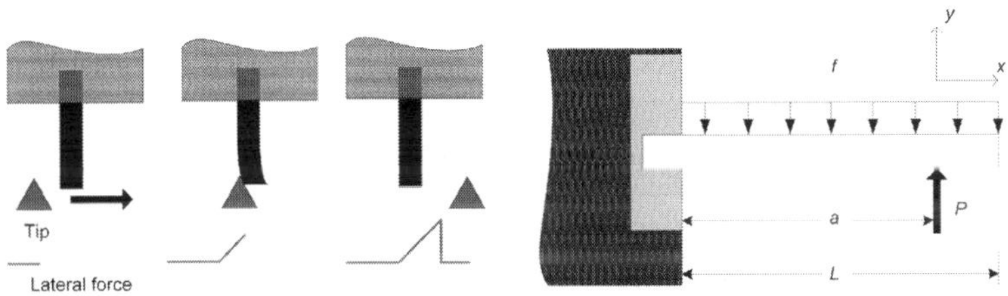

Figure 2.3 Left: Schematic of beam bending with the tip of an atomic force microscope (AFM). Right: Schematic of a pinned beam with a free end. The beam of length L is subjected to a point load P at $x = a$ and to a distributed friction force f. Adapted from Ref. [24].

corresponding linear increase in the lateral force with deflection. The right panel shows the system after the tip has passed over the beam. The lateral force drops to its initial value (left), and the beam snaps back to its undeflected, equilibrium position.

The measurement was done for six MWNTs of varying diameters. Nanotubes were deformed laterally by the AFM tip, until at a certain deformation the tip would pass over the tube, allowing the tube to snap back to its relaxed position. During measurements, lateral force–distance curves were acquired at different positions along the chosen beam. By modeling the nanotube as a beam and calculating its lateral force constant, one obtains

$$\frac{df}{dy} = \frac{3\pi d^4}{64x^3} Y,$$

where Y is Young's modulus of the beam; d its diameter; and x the position along the beam (see the right panel of Fig. 2.3). The resulting averaged value of Young's modulus was 1.28 ± 0.6 TPa, which was insensitive to the nanotube diameter.[a]

2.5 Doubly Clamped Deflection

Another possible AFT-based approach [14, 26] to probe the mechanical properties of carbon nanotubes consists of depositing bundles of SWNTs [14] or individual MWNTs [26] on an ultrafiltration membrane. In this approach, nanotubes are deposited on a flat

[a]For larger deformations, discontinuities in force–distance curves were observed, attributed the to buckling of nanotubes.

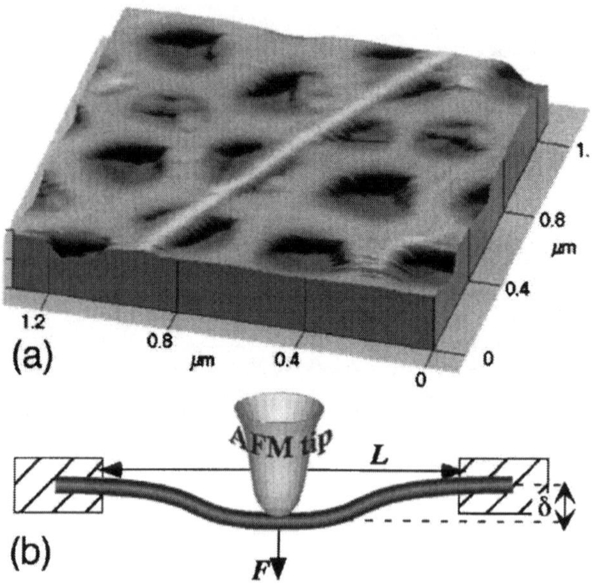

Figure 2.4 (a) AFM image of an SWNT rope adhered to the polished alumina ultrafiltration membrane, with a portion bridging a pore of the membrane. (b) Schematic of the measurement: the AFM is used to apply a load to the suspended nanotube and to determine directly the resulting deflection. Reprinted with permission from Salvetat, J.-P., Andrew, G., Briggs, D., Bonard, J.-M., Bacsa, R. R., Kulik, A. J., Stökli, T., Burnham, N. A., and Forró, L., *Phys. Rev. Lett.*, **82**(5), 944–947, 1999. Copyright (1999) by the American Physical Society.

alumina (Al_2O_3) ultrafiltration membrane that was endowed with a regular pattern of holes having a diameter of approximately 0.4 μm. When a nanotube (or its bundle) was found lying across one such hole (Fig. 2.4(a)), the AFM tip was used to apply a load on the suspended length of the tube (Fig. 2.4(b)), measuring the restoring force as a function of the deflection. Carbon nanotubes are assumed to stick to the surface due to van der Waals interaction so that they could not slide on the surface when the load was applied. As a result, the deflection versus force measurement allows the estimation of Young's modulus.

Measurements were carried out on both MWNTs grown by the arc-discharge method and those by catalytically grown method [26];

the latter MWNTs were structurally much less perfect than the first ones. The deflection δ of an isotropic beam as a function of the applied load is known from small deformation theory to be given by

$$\delta = \frac{FL^3}{\alpha YI}, \tag{2.7}$$

where F is the applied force, L is the suspended length, Y is the elastic modulus, I is the moment of inertia (or the second moment of area of the beam), and $\alpha = 192$ for a clamped beam. For a hollow cylinder,

$$I = \frac{\pi(a^4 - b^4)}{4}, \tag{2.8}$$

a and b being, respectively, the outer and inner radii. The measured value of Young's modulus ranged from an average of 0.8 TPa for arc-discharge grown nanotubes to average of 27 GPa for catalytically grown nanotubes. The latter values are markedly lower than those previously obtained [16, 24], which may due either to a higher concentration of structural defects or to a partial breakdown of the assumption of pinning of the nanotubes to the substrate.

In the case of bundles of SWNTs [14], the measurement is more difficult partly because of the small diameters and partly because of the fact that individual SWNTs comprising the bundle can easily slide to each other due to the weak van der Waals interaction that holds the individual tubes together. Owing to the weak interaction, bundles behave as a loose assembly of individual tubes rather than as a compact beam. When SWNT bundles are exposed to forces, they behave as an assembly, and the contribution of the shearing between carbon nanotubes becomes larger as the bundle diameter increases [14, 27]. The bundles have an extremely low shear modulus in the order of 1–10 GPa. As a result, SWNT bundles have a very low shear modulus ($\sim$1 GPa), implying that nanotubes in a bundle easily slide against each other.

This low shear modulus is a serious weakness of carbon nanotubes in practical use. In fact, the mechanical strength of a macroscopic carbon nanotube fiber is strongly dominated by their shear modulus and not by their Young's modulus, and consequently, the distinctive high Young's modulus of individual SWNTs is not

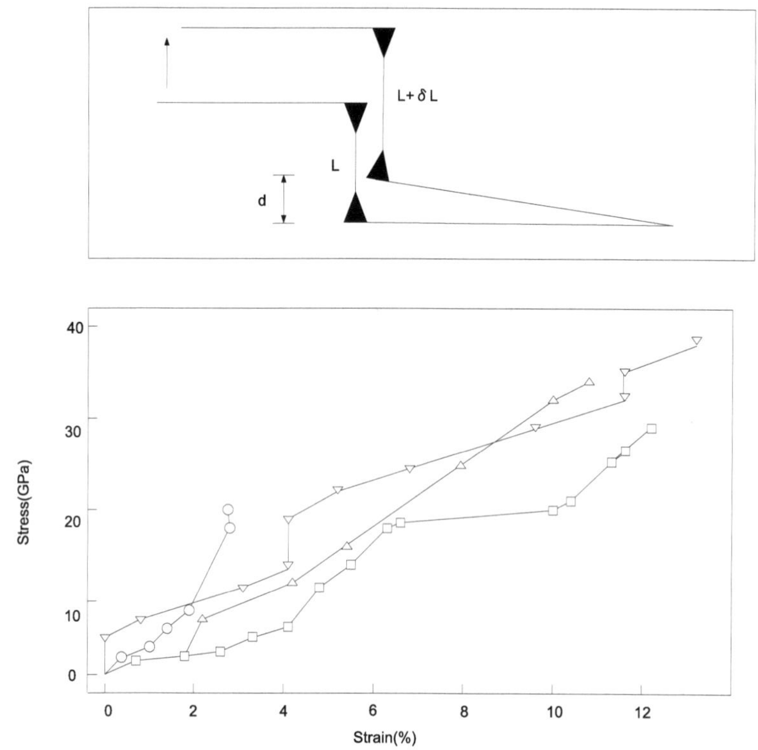

Figure 2.5 Top: Schematic showing the principle of the tensile-loading experiment. When the top cantilever is driven upward, the lower cantilever is bent upward by a distance d, while the nanotube is stretched from its initial length of L to $L + \delta L$ because of the force exerted on it. Bottom: Plot of stress versus strain curves for individual MWNTs. The $Y = 954$ GPa was obtained from a linear fit to the data points of open circles, and approximately 270–470 GPa from each of other curves. Adapted from Ref. [29].

exploitable. One way of solving this problem could be the creation of stable bonds between neighboring tubes. In order to avoid the slide process, cross-links between carbon nanotubes have to be generated. Electron beam irradiation [28] is one of the possible routes as discussed in Chapter 10.

2.6 Tensile Loading

The tensile loading experiment is one of the ordinary techniques for macroscopic elastic materials. To apply this technique to a carbon nanotube, it is connected to two opposing AFM cantilevers as shown in Fig. 2.5. The one cantilever is "compliant" and the other should be "stiff" such that the ratio of the cantilever spring constants is typically greater than 10. By recording the whole tensile-loading experiment, both the deflection of the soft cantilever (that determines the force applied on the nanotube) and the change in the nanotube length were simultaneously obtained.

In the measurement reported in Ref. [29], one AFM tip was integrated with a rigid cantilever with a spring constant above 20 N/m and the other was compliant (0.1 N/m). Once the carbon nanotube to be tested is properly attached to the AFM tips, the stiff (top) cantilever is driven upward, resulting in the lower cantilever being bent upward by a distance d. The nanotube is stretched from its initial length of L to $L + \delta L$ due to the force exerted by the AFM tips. The force is calculated as kd, where k is the force constant to the lower cantilever. As a result, the force applied on the nanotube, kd, and the strain of the nanotube, $\delta L/L$, were simultaneously measured, which provides the stress-strain curves. Young's modulus of MWNTs obtained in this fashion ranged from 270 to 950 GPa [29]. This nano-manipulation technique can be also applied to SWNT bundles [30], revealing 320–1470 GPa.

Chapter 3

Atomistic Modeling

Seeing more beauty in something more difficult is a common evil of men.

— René Descartes, in "Rules for the Direction of the Mind"

3.1 Introduction

In the previous chapter, we introduced Young's modulus measurement methods from experimental point of view. However, there is still few appropriate measuring techniques at the nanoscale order and it is really hard to obtain some reliable data from indirect measurements. Thus, the theoretical techniques play an important role in investigating the mechanical properties of carbon nanotubes.

In this chapter, we survey the theoretical methodology of "atomistic" techniques. The atomic simulations of carbon nanotubes are mainly based on: (i) the classical molecular dynamics (MD) with empirical interatomic potentials [31, 32]; (ii) the tight-binding and other semi-empirical methods [33]; and (iii) the first-principle quantum mechanics (QM) calculations [35].

Elastic and Plastic Deformation of Carbon Nanotubes
Hiroyuki Shima and Motohiro Sato

ISBN 978-981-4310-75-8 (Hardcover), 978-981-4364-15-7 (eBook)
www.panstanford.com

3.2 Methodology

3.2.1 Ab initio *Method*

Ab initio method (or often called the first-principle method) is a method of calculating atomic and molecular structure directly from the first principles of quantum mechanics, without using quantities derived from experiment as parameters. It can be used, for example, to determine the bond lengths and bond angles of molecules by calculating the total energy of the molecule for a variety of molecular geometries and finding which conformation has the lowest energy.

The central theme of *ab initio* methods is to obtain accurate solutions to the Schrödinger equation. In general, the state of a particle is defined by a wave function ψ based on the well-known wave-particle duality. The Schrödinger equation is

$$H\psi = E\psi,$$

where H is the Hamiltonian operator of the quantum mechanical system, and ψ is the energy eigenfunction corresponding to the energy eigenvalue E.

Since the pristine *ab initio* method is too expensive and time consuming, it is in practice incorporated with some approximations such as the density functional approach. The density functional theory (DFT) is based on the fact that the ground-state electronic energy is a unique functional of the electronic density $\rho(\boldsymbol{r})$, instead of many-body interacting wave functions [36]. In typical cases, the electronic energy is expressed as

$$E = T[\rho(\boldsymbol{r})] + \int \frac{[\rho(\boldsymbol{r})\rho(\boldsymbol{r}')]}{|\boldsymbol{r}-\boldsymbol{r}'|} d\boldsymbol{r}d\boldsymbol{r}' + \int V_{\mathrm{N}}(\boldsymbol{r})d\boldsymbol{r} + E_{\mathrm{XC}}[\rho(\boldsymbol{r})]. \quad (3.1)$$

In Eq. (3.1), $T[\rho(\boldsymbol{r})]$ is the kinetic energy and is a function of the electron density, the second term represents the electrostatic potential, the third term denotes the contribution from the nuclei potential, and the last term is the exchange-correlation functional. By introducing the variational principle, we obtain a one-electron Schrödinger equation (also called the Kohn–Sham equation) for the Kohn–Sham wavefunction ψ_λ [37]

$$\left\{-\frac{1}{2}\nabla^2 + v(\boldsymbol{r}) + \int \frac{\rho(\boldsymbol{r}')}{|\boldsymbol{r}-\boldsymbol{r}'|} d\boldsymbol{r}' + \mu_{\mathrm{XC}}[\rho](\boldsymbol{r})\right\} \psi_\lambda = \epsilon_\lambda \psi_\lambda,$$

where

$$\mu_{XC}[\rho](\mathbf{r}) = \frac{\partial E_{XC}[\rho(\mathbf{r})]}{\partial \rho}.$$

The problem is reduced to obtaining the solutions to systems of one-electron equations. Once ψ_λ and ϵ_λ are solved, the total energy and the corresponding electron state can be obtained from Eq. (3.1).[a]

3.2.2 *Tight-Binding Method*

The tight-binding method is a class of a one-electron model, allowing the calculation of electronic states using a set of wave functions for isolated atoms located at each atomic site. The name "tight binding" refers to the fact that the electrons in this model should be regarded as tightly bound to the atom to which they belong; therefore, the wave function of the electron will be rather similar to the atomic orbital of the free atom where it belongs to.

The theoretical framework of the tight-binding model was originated by Slater and Koster [40]. The advantage of the tight-binding method is that it can handle a much larger system than the *ab initio* method while maintaining better accuracy than MD simulation. In the tight-binding method, a linear combination of atomic orbitals is adopted in the wave function. Although the exact forms of the basis are not known, the Hamiltonian matrix can be parameterized, and the total energy and electronic eigenvalues can be deduced from the Hamiltonian matrix. The interatomic forces are evaluated using the Hellmann–Feynman theorem and the rest of the procedure is almost identical to the MD simulation. For this reason, sometimes the tight-binding method is also referred to as the tight-binding MD method.

---

[a]The solution procedure requires an iterative diagonalization process, which in general involves $O(N^3)$ order of computation. Major improvements have been made using the Car-Parrinello molecular dynamics method [38] and conjugate gradient method [39]. The Car-Parrinello method reduces the order from $O(N^3)$ to $O(N^2)$. As shown in [39], the conjugate gradient method can even be more efficient.

3.2.3 *Molecular Dynamics Method*

An MD method was developed to calculate positions and velocities of constituent atoms (or molecules) in a material considered, owing to which the overall dynamics of the material can be reproduced in computer simulations. In the classical MD method, the objective is to solve the governing equations of particle dynamics based on Newton's second law, i.e.,

$$m_{\mathrm{i}}\frac{d^2\boldsymbol{r}_{\mathrm{i}}}{dt^2} = -\nabla V,$$

in which m_{i} and $\boldsymbol{r}_{\mathrm{i}}$ are the mass and spatial coordinates of the i-th atom, respectively, V is the empirical potential for the system, and ∇ denotes the spatial gradient. Due to the small time scale involved, explicit integration algorithms such as the Verlet method [41] and other high-order methods are commonly used to ensure high-order accuracy.

3.3 Comparison: Quantum or Classical?

Having known in Section 3.2 that many computational methods have been developed to analyze the structural properties of materials, we may have a question as to how the differences are involved between them. At the most detailed level, quantum mechanical approaches such as the *ab initio* or the high-level DFT-based calculations lead the precise solutions of the electric charge density distribution and electrostatic energy. Thus, they provide accurate solutions for atomic and molecular structural deformation and fracture, as well as any chemical reaction that may occur [35, 42, 43].

Despite of the numerical accuracy, the quantum mechanical calculation is computationally expensive and only effective to small systems containing several hundreds of atoms. This means a limited availability of the method for carbon nanotubes, only up to several nanometers in length to be dealt with. In order to reduce the computational cost and still keep the quantum mechanical characteristics, a semi-empirical method such as tight binding (TB) was developed to solve the molecular orbital functions by replacing complex integrals with simpler empirical parameters and functions

[44]. The semi-empirical/TB approach can be applied to larger systems (with the carbon nanotube length less than 100 nm) [45, 46].

With the development of a better force field, i.e., less empirical [47] and numerical algorithms, the classical MD simulations have been shown to play an important role in revealing the constitutive mechanisms of the carbon nanotubes, including elastic deformation under tension, bending, torsion, radial pressure, and buckling instabilities caused by compression. Moreover, the geometrical imperfections (which arise during the equilibrium optimization of the atomic structure) and defects (e.g., atomic vacancies or the Stone–Wales defect where the bonds rearrange into pentagon or heptagon pairs) have been shown to play a critical role in the mechanical integrity [48–50].

It should be noted that classical MD simulations alone cannot be used to study the fracture of the carbon nanotubes. For this, the Tersoff and Brenner potentials-based MD or multiscale coupling between the TB potentials and MD are needed (in order to simulate larger systems), as discussed in detail in the next section .

3.4 Interatomic Potential

3.4.1 *Tersoff and First-Generation Brenner Potential*

In general, carbon–carbon bond interactions are described by either analytic potential energy expression or semi-empirical structure methods, or obtained from a total-energy first-principle calculations. Although the latter calculation does not need any assumptions of functional forms and parameter fitting, it is relatively computer time intensive. The approaches using the classical potential energy expression have, therefore, to model large systems that have a huge number of carbon atoms. Furthermore, closed-form expressions in these potentials enable analytic evaluation of derivatives as required in computations of stresses and forces although the accuracy of the results is limited by the extent of applicability of a given potential to a certain atomic system. Here we show some well-known Tersoff and Brenner interatomic potential energy expressions for carbons. Such potentials can accurately describe the bonding energies and elastic properties of a wide range of graphite phases of carbon as well as diamonds and hydrocarbons.

The general analytical expression using the interatomic potential energy was originally presented by Abell with empirical chemical pseudopotential theory [51]. In this article, the general expression was derived as a sum of nearest neighbor pair interaction that are moderated by the local atomic environment. After that, the analytical parameterized potential energy function based on the Abell expression was introduced by Tersoff [52–55]. Two parameterizations of a expression with Morse-type potential were given to be fit to the energetics of a relatively large number of molecules. It is the well-known Tersoff potential.

Later in 1990, Brenner pointed out several problems of the Abell–Tersoff formalism in his article [56]. He insisted that the first parameterization provided a good description of carbon–carbon bond lengths stretching force constants were evaluated too small, while the second parameterization gives force constants that are in better agreement with experiment but produces bond lengths for double and triple bonds that are too long. From this viewpoint, Brenner introduced a new empirical potential energy expression based on Tersoff's covalent-bonding formalism with additional terms that correct for an inherent overbinding of radicals and that include non-local effects [56]. It is called the first-generation Brenner potential.

The potential expression between pairs of nearest-neighbor atoms i and j takes the form [56]

$$V_{\mathrm{ij}} = V^{R}(r_{\mathrm{ij}}) - \bar{b}_{\mathrm{ij}} V^{A}(r_{\mathrm{ij}}), \tag{3.2}$$

where r_{ij} is the distance between atoms i and j, $V^R(r)$ and $V^A(r)$ are the repulsive and attractive pair-additive interaction terms, and $\bar{b}_{\mathrm{ij}}$ is the empirical bond order function describing the bonding environment. In the first-generation Brenner potential, $V^R(r)$ and $V^A(r)$ are given with an optional cutoff function to restrict the range of the potential $f^c(r)$ as

$$V^{R}(r) = f^{c}(r)\frac{D_{\mathrm{ij}}^{(e)}}{S_{\mathrm{ij}} - 1}\exp\left[-\sqrt{2S_{\mathrm{ij}}}\beta_{\mathrm{ij}}(r - R_{\mathrm{ij}}^{(e)})\right], \tag{3.3}$$

and

$$V^A(r) = f^c(r)\frac{D_{\text{ij}}^{(e)}S_{\text{ij}}}{S_{\text{ij}}-1}\exp\left[-\sqrt{2/S_{\text{ij}}}\beta_{\text{ij}}(r - R_{\text{ij}}^{(e)})\right]. \tag{3.4}$$

respectively [56]. Note that these pair terms reduce to the usual Morse-type potential used in Tersoff formalism [52–55] if $S_{\text{ij}} = 2$ in Eqs. (3.3) and (3.4). Moreover, the equilibrium distance $R_{\text{ij}}^{(e)}$, well depth $D_{\text{ij}}^{(e)}$ and β_{ij} are equal to the usual Morse parameters, which depends on the value of S_{ij}. The empirical bond order function $\bar{b}_{\text{ij}}$ is given by

$$\bar{b}_{\text{ij}} = \frac{1}{2}(b_{\text{ij}}^{\alpha-\pi} + b_{\text{ji}}^{\alpha-\pi}) + \Pi_{\text{ij}}^{RC}, \tag{3.5}$$

in which the first term of the left-hand side in the above equation is the average of terms associated with each atom in a bond, which values for the functions $b_{\text{ij}}^{\alpha-\pi}$ and $b_{\text{ji}}^{\alpha-\pi}$ depend on the local coordination and bond angles, for atoms i and j, respectively. The second term Π_{ij}^{RC} in Eq. (3.5) is dependent of whether a bond between atoms i and j has radical character and is part of a conjugated system. Thus, only second nearest neighbor interactions are incorporated in the Tersoff and first-generation Brenner potential. Hence, the functions is quickly evaluated and can be used in large scale MD simulations.

3.4.2 *Second-Generation Brenner Potential*

As shown in the previous subsection, Tersoff [52] and first-generation Brenner [56] potentials take into account many-body interactions only up to the second nearest neighbors for carbon atoms. In addition, Morse-type functions used in such potentials is too restrictive to fit equilibrium distances, energies and force constants for carbon–carbon bond simultaneously. In contrast to these facts, in the "second-generation" Brenner potential, effects of the third nearest-neighbors via a bond-order term associated with the dihedral angles are also included and modified pair-additive interaction functions are introduced [57]. Thus, the new second-generation Brenner potential function gives a much improved description of bond energies, bond lengths, and force constants for carbon–carbon bonds as compared with the previous-generation potentials. We show below the properties of the second-generation Brenner potential with a contrast to the Tersoff and first-generation Brenner potential.

In the Tersoff and the first-generation Brenner potential, the pair-additive interaction terms were expressed by Morse-type terms. On the other hand, the following functional forms for the pair terms $V^R(r)$ and $V^A(r)$ are used in the second-generation Brenner potential [57]:

$$V^R(r) = f^c(r)\left(1 + \frac{Q}{r}\right) A \exp(-\alpha r), \tag{3.6}$$

and

$$V^A(r) = f^c(r) \sum_{n=1,3} B_{\mathrm{n}} \exp(-\beta_{\mathrm{n}} r), \tag{3.7}$$

respectively. Following the earlier hydrocarbon bonding expression, the empirical bond order function $\bar{b}_{\mathrm{ij}}$ for the second-generation Brenner potential is written as a sum of terms:

$$\bar{b}_{\mathrm{ij}} = \frac{1}{2}(b_{\mathrm{ij}}^{\alpha-\pi} + b_{\mathrm{ji}}^{\alpha-\pi}) + \Pi_{\mathrm{ij}}^{RC} + b_{\mathrm{ij}}^{DH}, \tag{3.8}$$

where the first and second terms of the left-hand side in Eq. (3.8) corresponds to Eq. (3.5), i.e. the first-generation Brenner potential [56], and the third term b_{ij}^{DH} express a function of the dihedral angles for carbon–carbon double bonds, which takes the form [57]

$$b_{\mathrm{ij}}^{DH} = \frac{T_0}{2} \sum_{k,l(\neq i,j)} (1 - \cos^2 \Theta_{\mathrm{ijkl}}) f_{\mathrm{ik}}^c(r_{\mathrm{ik}}) f_{\mathrm{jl}}^c(r_{\mathrm{ik}}). \tag{3.9}$$

The top of the left and right panels of Fig 3.1 shows the plots of the bond energies and the force constants determined by the second-generation Brenner potential. Those obtained from the first-generation Brenner potential are also plotted in the bottom graphs of Fig. 3.1 for reference. (Note that the difference between potential 1 and 2 comes from their parameterization.) We can clearly see from Fig 3.1 that the Morse-type functions used in the earlier-generation hydrocarbon potentials are unable to describe bond energies, distances, and force constants simultaneously for C–C bonds. On the other hand, the functions developed in the second-generation Brenner potential are able to express each of these properties relatively accurately.

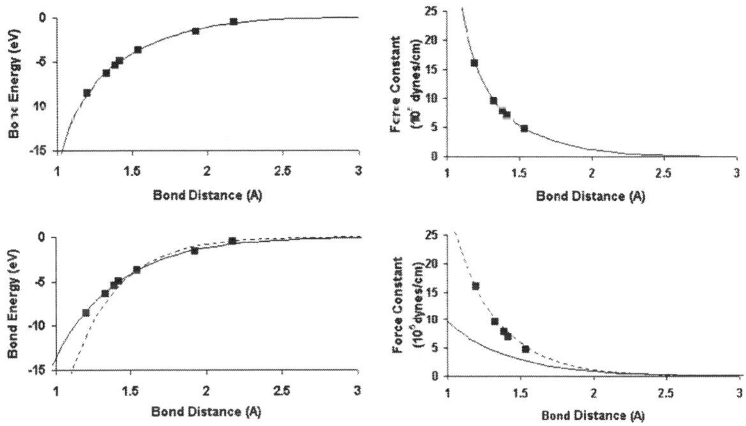

Figure 3.1 Plots of pair term properties. Top left and right panels: plots of bond energy and force constant, respectively, versus bond distance for the pair terms developed in the second-generation Brenner potential. Bottom panels: corresponding properties for potential 1 (solid lines) and potential 2 (dotted lines) in the first-generation Brenner potential. Reprinted with permission from Brenner, D. W., Shenderova, O. A., Harrison, J. A., Stuart, S. J., Ni, B., and Sinnott, S. B. (2002). A second-generation reactive empirical bond order (REBO) potential energy expression for hydrocarbons, *J. Phys. Condens. Mat.*, **14**, pp. 783–802.

3.5 Young's Modulus Prediction

Accurate numerical simulations using tight-binding molecular dynamics methods and *ab initio* density functional total energy calculations with realistic strain rate, temperature dependence, and carbon nanotube sizes have provided more realistic values of 1 TPa as Young's modulus and 5 to 10 % elastic limit of the tensile strain before failure [58]. In general, it has been understood that the disadvantages of using the atomistic-based techniques for evaluation of elastic properties are, for example,

(1) Their formulations are complicated,
(2) They are not suitable to model carbon nanotubes including huge amount of atoms,
(3) Computational costs become expensive, and so on.

Nevertheless, these methods are really effective especially for relatively small scale problems because of their correctness. We can find many theoretical research studies by the atomistic-based approaches to investigate the variation of Young's modulus and Poisson's ratio as a function of the carbon nanotube diameter. In the following, we provide an overview of numerical Young's modulus measurements of carbon nanotube using atomistic-based approach, such as *ab initio* approaches, tight-binding methods and MD.

Contributions to the good elastic-mechanical characteristics of carbon nanotubes come mainly from the strength of in-plane covalent C–C bonds in graphene sheet and facile out-of-plane deformation of the structure. For large-diameter or small-curvature carbon nanotubes, the modulus, strength, and stiffness should therefore be comparable with the in-plane modulus and strength of graphene sheet. In a tubular geometry with small-diameter, however, the elastic strain energies are also affected by the intrinsic curvature of the surface. Using the Tersoff and Brenner potentials introduced before, it was shown that the elastic energy of an SWNT scales as R^{-2}, where R is the radius of the tube [59]. This is similar to the results deduced from the continuum elastic theory [60]. For axial strains, Young's modulus Y of an SWNT is defined as

$$Y = \frac{1}{V}\frac{\partial^2 E}{\partial \varepsilon^2}, \tag{3.10}$$

where E is the strain energy and V is the volume of the nanotube. Initial computation studies [32] reported the value of Young's modulus to be as high as 5.5 TPa. This was mainly due to a very small value of carbon nanotube wall thickness ($\sim$0.06 nm) used in this study [32]. It turns out that this was partly because of the attempts to use continuum theory to describe equivalence between a shell model and the atomistic descriptions of the elastic properties of carbon nanotubes. In many later works, this discrepancy [32] has been corrected. The van der Waals radius of carbon atoms can be used to define the spatial extent of the atoms, and the single-wall thickness comes to be about 0.34 nm, which corresponds to the graphene sheet thickness. Using an empirical force constant model, Lu found that Young's modulus of an SWNT is around 970 GPa, which is close to that of an in-plane modulus of a graphene sheet, and is independent of tube diameter and chirality [61]. A better description for the interatomic forces through a non-orthogonal tight-binding methods was used to find Young's modulus to be around 1.2 TPa [33], which is larger than that of in-plane graphene sheet value and is slightly dependent on the tube size for small diameter ($D < 1.2$ nm) carbon nanotubes.

To summarize, we found that the most widespread value for Young's modulus is around 1 TPa by taking a value of 0.34 nm for the graphene sheet thickness, although the numerical estimation of Young's modulus is dependent of the carbon nanotube diameter. In other words, it can also be said that the value of (Young's modulus)×(wall thickness) should roughly be evaluated as about 0.34 TPa·nm. We will see below more examples of some related numerical research studies.

3.6 Via *ab initio* Approach

As described before, the *ab initio* simulations that are potential-free approaches tend to be restricted to systems with small number of atoms because of their computational expense. Nevertheless, these approaches have the powerful advantage of being highly transferable and quantitatively reliable. In these techniques, the force on atoms is found from electric structure calculation as a progressive task.

Sánchez-Portal *et al.* carried out the *ab initio* calculations using an implementation of pseudopotential-density-functional theory [35]. In this research, an extensive study of the vibrational and mechanical properties of SWNTs was conducted, in both isolated and bundle geometries, employing DFT calculations [62] code. As shown in Fig. 3.2. The authors obtained values of Young's modulus of around 1 TPa for SWNTs in an isolated geometry, whereas in a bundle geometry, this value came down to 0.5 TPa for a bundle of (10,10) nanotubes. The results of Lu [61], Hernández *et al.* [33, 63], and Sánchez-Portal *et al.* [35] essentially agree on the finding that Young's modulus is insensitive to the diameter and chiral angle of nanotubes except for very low diameters and provide similar values of Young's modulus when the same convention is used to define the thickness of the nanotube shell ($\delta = 0.34$ nm). These

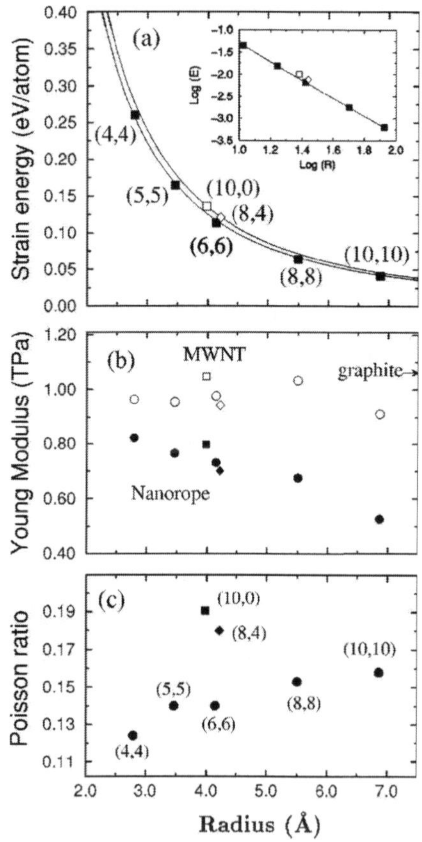

Figure 3.2 (a) Strain energy versus tube radius. Solid line drawn across the (n,n) data corresponds to a least-squares fit to the C/r^2 behavior (where r is the radius of the tube and C is a constant that depends on Young's modulus Y and thickness h of the wall in a model tube: $C = Yh^3a/24$). The two C/r^2 functions passing through the (8,4) and (10,0) data are also shown (in this scale they appear as one line). The $r^{-\alpha}$ behavior is clearly shown in the inset. The value obtained from the logarithmic fit is 2.05 ± 0.02. (b) Young's modulus versus tube radius. Open symbols for the multi-wall geometry, and solid symbols for the SWNT crystalline-rope configuration. The experimental value of the C_{11} elastic constant of graphite is also shown. (c) Poisson ratio versus tube radius. Reprinted with permission from Sánchez-Portal, D., Artacho, E., Soler, J. M., Rubio, A., and Ordejón, P., *Phys. Rev. B*, **59**, 12678–12688, 1998. Copyright (1998) by the American Physical Society.

theoretical results are also in good agreement with the experimental measurements of Wong *et al.* [24] and those of Krishnan *et al.* [17] for MWNTs and SWNTs.

More recently, Chandraseker and Mukherjee demonstrated the stress–strain relationship and the elastic moduli for SWNT by a fully *ab initio* density functional theory approach as well as a quasicontinuum method that involves the use of the empirical Tersoff–Brenner potential [64]. In this article, the above two different approaches are compared and the advantages and limitations are discussed. The authors pointed out that quasicontinuum analysis are best suited to obtain efficient qualitative predictions, while *ab initio* simulations are necessary to extract quantitatively accurate parameter values for use in "suitable" continuum models of atomic systems.

3.7 Via Tight-Binding Approach

Hernández *et al.* investigated the mechanical properties of SWNTs of different compositions, including C, BN, BC_2N, C_3N_4, and BC_3, which were obtained employing a non-orthogonal tight-binding model [44] parameterized on density functional theory (DFT) [33, 63]. This study confirmed that among the different compositions considered, carbon nanotubes have the highest Young's modulus, with a value of 1.26 TPa for the widest tubes, although this value was largely insensitive to the nanotube diameter except for very narrow nanotubes, where the high curvature resulted in a weakening of the chemical bonds and consequently in a reduction of Young's modulus. BN nanotubes were calculated to have a Young's modulus of 0.8 TPa, which, although lower than that of carbon nanotubes, is still very high (higher than that of SiC nanorods, for instance).

Figure 3.3 shows Young's modulus of a variety of carbon and non-carbon nanotubes as a function of the tube diameter [33]. It has been suggested in this research that by investigating the value $\partial^2 E/\partial \varepsilon^2$, instead of Young's modulus, the ambiguity of thickness of a carbon nanotube wall can be completely avoided. However, the value of 0.34 nm for the wall thickness of an SWNT gives atomistic simulation-based results of Young's modulus that are in broad agreement with experimental results so far. Using a non-

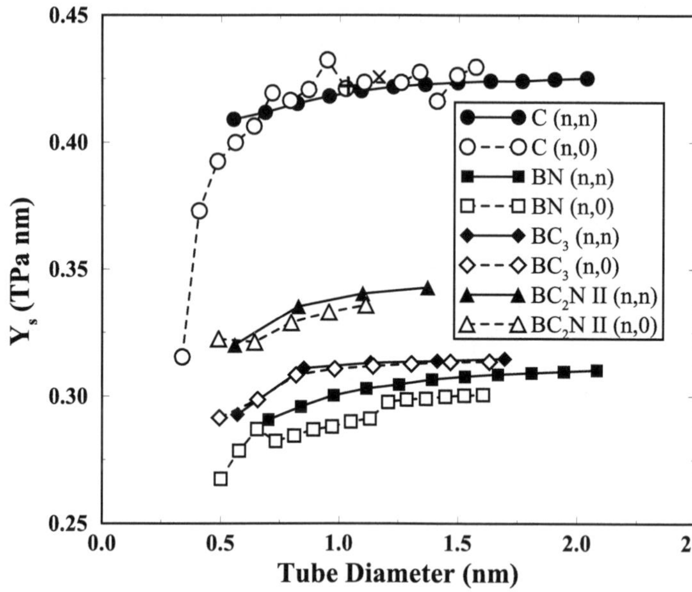

Figure 3.3 Young's modulus as a function of the tube diameter for C, BN, BC_3 and BC_2N, as calculated from tight-binding simulation. Result obtained for (n, n) nanotubes (filled symbols), $(n, 0)$ nanotubes (empty symbols) and also for C (10, 5) (+) and (10, 7) (×) are shown. Reprinted with permission from Hernández, E., Goze, C., Bernier, P., and Rubio, A., *Phys. Rev. Lett.*, **80**, 4502–4505, 1998. Copyright (1998) by the American Physical Society.

orthogonal tight-binding, molecular dynamics, and DFT methods, the axial compression of SWNTs and boron nitride nanotubes were carried out recently. Young's modulus of SWNTs was found to be 1.2 TPa and that of boron-nitride nanotubes was 80 % of that of equivalent carbon nanotubes [34]. These results are in qualitative and quantitative agreement with the DFT results and general experimental observations of Sánchez-Portal *et al.* [35] known thus far.

3.8 Via MD Approach

In molecular dynamics (MD) modeling, parameters in the potentials should be usually fit experimentally or from *ab initio* simulation results corresponding to specific physical criteria. For this reason, these approaches may not be applicable well to other systems. These are really suitable, however, for analyses of large atomic system because of their computational efficiency.

To our knowledge, the first theoretical study on the energetics and elastic properties of carbon nanotubes was that due to Robertson *et al.* [59] published in 1992. In this article, these authors performed calculations on models of carbon nanotubes using two different empirical potentials, namely those of Tersoff [52] and Brenner [56], introduced in the previous section. For a selected subset of nanotubes calculations were also performed employing first-principle techniques as a check on the reliability of the empirical potentials. This study established that the strain energy, defined as the energy difference per atom of a given nanotube and an infinite flat graphene sheet, scales as the inverse of the squared radius of the nanotubes, in agreement with what had been predicted by classical elasticity theory [60]. This is so even down to very narrow nanotube radii.

Concerning Young's modulus, the calculations show that in the limit of large radii, the stiffness of the tube corresponds directly to the C_{11} elastic constant of graphite calculated with the same model. The accepted experimental value of C_{11} of graphite is 1.06 TPa [65, 66], and the empirical potential of Brenner [56] gave a value in close agreement with this result, while that of Tersoff [52] slightly overestimated it.

Another interesting result to be derived from this study was that both empirical models predicted that Young's modulus should decrease for very narrow (radius $\leq$ 4 Å) nanotubes. This result is physically intuitive, because as the curvature strain is increased, the C–C bonds are weakened, and this effect should be reflected in the mechanical properties.

After the above research by Robertson *et al.*, some notable molecular dynamics based approaches using the empirical potentials can be found. For example, Yakobson *et al.* [32] performed a series of atomistic molecular dynamics simulations of nanotubes subject to large compressive strains, and to bending or torsional distortions using the Brenner potential [56]. For nanotubes under compressive strain, it was found that the tubes began to buckled at critical strain values at which the energy of the system displayed a slope discontinuity. The appearance of these buckles was rationalized in terms of a continuum shell model with appropriately fitted parameters of flexural rigidity and in-plane stiffness. This model was also able to account for the buckling observed under bending and torsional deformations. The atomistic simulations of this study confirmed that SWNTs could sustain large structural distortions without any indication of bond breaking or rearrangement of the bonding pattern, demonstrating the high elasticity of nanotubes. The experimental observation [67] of similarly distorted nanotubes served as a confirmation of the conclusions derived from this study.

A detailed study of the mechanical properties of nanotubes was carried out by Lu [61]. In this work, Young's modulus, the Poisson ratio (defined as minus the radial strain divided by the axial strain) and other elastic constants were calculated for a series of SWNTs, MWNTs, and crystalline ropes of SWNTs, using an empirical force field fitted to reproduce the vibrational frequencies and elastic constants of graphite. It was found that Young's modulus of SWNTs was insensitive to the nanotube diameter and chiral angle. The lack of sensitivity to the nanotube diameter, even for narrow nanotubes, was in contradiction with the earlier finding of Robertson *et al.* [59]. The reason for this disagreement can be ultimately traced to the simplicity of the model used by Lu, which included only harmonic springs to represent C–C bonds and pairwise Lennard-Jones type potentials to model the interlayer interaction in MWNTs. In the case of bundles of SWNTs, it was found that Young's modulus decreases as the diameter of the nanotubes forming the bundle is increased. However, this does not reflect a weakening of the C–C bonds in these bundles but simply is a consequence of the fact that unit-cell volume grows more rapidly than the number of atoms. Therefore, the decreasing value of Y is a consequence of the peculiar geometry of the system. The values of Y reported by Lu varied between 1.11 TPa for MWNTs and 0.97 TPa for isolated SWNTs.

Cornwell and Wille obtained Young's modulus of SWNTs under axial compression by the MD calculation [68]. In this research, the authors approximated Young's modulus Y using a least squares fit as

$$Y = \frac{4296\ [\text{GPa Å}]}{r\ [\text{Å}]} + 8.24[\text{GPa}].$$

The values obtained from this formulation agree reasonably well with a methods for approximating Young's modulus developed by Ruoff and Lorents [84].

3.8.1 *Dependence on Diameter and Chirality*

The hexagonal symmetry of a graphene sheet ensures a perfect isotropy of its in-plane elastic behavior. As the sheet is rolled seamlessly into a tube, the bending strain induces a slight out-of-plane distortion of the C–C bonds, resulting in a hybridization of the σ bonds, which increases as the diameter decreases. The hexagonal symmetry is thus broken so that its elastic behavior may vary slightly with the shell chirality. The dependence of the elastic strain energy to the diameter and chirality is an important issue especially for growth models.[a] Suppose that the energy difference between armchair and zigzag nanotube is much higher than $k_B T$, one may hope to find out a synthesis method producing only one chirality.

It seems clear that calculating the diameter and chirality dependence of Young's modulus goes beyond the isotropic elastic shell approximation. Only with models taking explicitly into account the bound deformation due to curvature, one may expect to find anisotropic effects in elastic properties. Different approaches are reported in the literature, which often bring different results (sometimes contradictory), and discerning the origin of these differences is often a case for the specialists.

The simplest approach to this problem is to consider that curving a C–C bond results in decreasing its stiffness (rotating a bound is much easier than stretching it). This implies a lowering of the nanotube axial stiffness as the diameter decreases. An armchair nanotube is thus expected to have a lower stiffness than a zigzag

[a]Carbon nanotubes are always produced with a distribution of chiralities over which there is no real control. From the mechanical point of view, strength and the Poisson ratio show the most pronounced chirality dependence, while Young's modulus is practically independent of chirality for nanotubes with diameters above 0.7 nm [33].

tube since all its C–C bonds are curved. Our assumption implicitly assumes that the in-plane isotropy is lost in the cylindrical geometry.

Do these basic (simplistic) arguments find a serious justification with sophisticated models? First-principal calculations [35] shows that bonds are inequivalent in length depending on their orientation to the tube axis, and that the bond length increases with decreasing tube radius, the greatest increase being for perpendicular bonds. This is a rigorous statement of our qualitative approach since bond stiffness and bond length are correlated. Indeed most calculations agree that Young's modulus slightly decreases with tube diameter [33, 69, 70, 83].

Chapter 4

Continuum Modeling

Every man hath his own particular point of view, and, at different times, sees the same objects in very different lights.
— Cesare Beccaria, in "On Crimes and Punishments"

4.1 Introduction

The atomistic modeling approaches such as the molecular dynamics, the tight-binding method, and the first-principle quantum mechanics, which are shown in the previous chapter, yield results that are in many cases explicit in nature. However, they become less practical, especially in problems involving a large number of carbon atoms (e.g., long MWNTs or nanotube bundles), or to compare with experiments that are conducted at a much longer time scale due to the limitation in time scales (typically, several nanoseconds or less) and length scales of the models. Moreover, for practical applications, phenomenological continuum-based material parameters such as Young's modulus and Poisson's ratio need to be defined and measured for the carbon nanotubes. Therefore, it is of fundamental value to develop reliable and computationally efficient

Elastic and Plastic Deformation of Carbon Nanotubes
Hiroyuki Shima and Motohiro Sato

ISBN 978-981-4310-75-8 (Hardcover), 978-981-4364-15-7 (eBook)
www.panstanford.com

continuum models for the nanotubes, which should closely duplicate the constitutive behaviors obtained from atomistic simulations [71].

The theory of elasticity [72, 73] provides a conceptual framework for understanding the mechanical properties of carbon nanotubes. In addition, the continuum-based modeling covers its difficulties and disadvantages for atomistic modeling because of its simplicity and efficiency. Nevertheless, the use of the continuum modeling in the nanoscale world requires some care. In fact, the question may arise whether the continuum elastic theory, which is mainly used for describing "macroscopic" structural mechanics, is applicable to "nano" structures. (Originally, the structural mechanics technique has thus far been developed in the field of engineering, for example, civil, mechanical, space, naval, and so on.) The most important quantity to be determined in the continuum elastic theory is Young's modulus Y, which is constant for a given isotropic material. The unusual fact with nanotubes is that Y is expected to vary slightly with the diameter, since curvature modifies C–C bond hybridization, hence the stiffness. That is why careful survey to check the validation of the continuum modeling in carbon nanotube analyses is needed.

When we use the continuum elastic theory to model carbon nanotube mechanics, the linkage between the modeling of the nanoscale molecular structure and the macroscopic structural elements is of great importance. In this case, C–C bonds or the nanotube itself is modeled as macroscopic structural elements such as bars, trusses, beams, curved shells, and cylindrical shells. In this chapter, we present some linkage modeling techniques between carbon nanotubes and continuum-based structural mechanics after brief introduction of the fundamentals of continuum elastic theory as well as the fundamental elasticity and basic structural elements, which is frequently used in structural analysis of carbon nanotubes.

4.2 Young's Modulus

The classical theory of elasticity [72, 73] defines a series of moduli (Young's modulus, bulk modulus, elastic constants, etc.) that characterize the response of a material under a perturbation; the perturbation causes a change in its shape and/or volume from that of its equilibrium configuration. This is well-known "Hooke's law," which applies only up to the proportional limit of the material. For example, Young's modulus, Y, which defines the response of the system when it is strained along a particular axis, is given by

$$Y = \frac{1}{V_{\text{eq}}}\left(\frac{\partial^2 E}{\partial \varepsilon^2}\right)_{\varepsilon=0}, \tag{4.1}$$

where E is the strain energy of the system, ε is the normal strain in the axial direction, which is defined by

$$\varepsilon = \frac{L - L_{\text{eq}}}{L_{\text{eq}}},$$

L being the length of the system along the strain axis, and L_{eq}, V_{eq} are the equilibrium length and volume of the system, respectively. Other elastic moduli are defined in similar ways.

Since the strain energy E is an extensive property that grows with the size of the system, the factor of V_{eq}^{-1} is introduced in Eq. (4.1) so as to make the Y intensive (i.e., size independent). However, the definition of V_{eq} can be problematic in nanoscaled objects, because there is no obvious way of defining the thickness of a shell that is only one atom thick. This difficulty has been frequently overcome by assigning an arbitrary thickness to the wall (e.g., the interlayer spacing in graphene, 0.34 nm), but there is no universally followed convention. Unlike that of a macroscopic tube, the "thickness" of a graphene wall seems to be a rather arbitrary notion because it has a meaning only in the framework of a continuum model. It is, however, an important quantity since it determines the moment of inertia of the nanotube, speaking in terms of continuous elasticity, and so the flexural behavior.

A number of papers have addressed this problem using different approaches [74, 79–82]. There are quite a number of publications containing seemingly contradictory results obtained from similar computer simulations, where the difference arises because a different convention in the definition of the elastic properties has been used, a situation that leads to some confusion as to what the actual value of, e.g., Young's modulus really is.[a]

---

[a]A number of calculations on nanotube elasticity have been performed since the discovery on 1991. They have resulted in some confusion that results from the different results reported by various authors. Yackobson *et al.*, reported $Y \simeq 5$ TPa and a thickness h of $h \simeq 0.07$ nm [35, 83–85].

4.3 Axial Rigidity for Bars and Trusses

In many engineering fields, axially loaded members are modeled as "bars" or "trusses." For such members, only normal stresses develop on sections perpendicular to the bar axis, and these members are not expected to resist bending and torsional stresses. Here we consider the deformation mechanics for a bar of uniform cross section. For linear elastic materials with Young's modulus Y, according to Hooke's law, we have normal strain ϵ_{x} and stress σ_{x} in the axial (x-) direction as follows:

$$\epsilon_{\mathrm{x}} = \frac{\sigma_{\mathrm{x}}}{Y} \tag{4.2}$$

$$\sigma_{\mathrm{x}} = \frac{P(x)}{A} \tag{4.3}$$

where $P(x)$ is the applied force in the axial direction and A is the cross-sectional area of the bar. In addition, the axial displacements $u(x)$ of the bar is, in essence, determined by solving a first-order differential

$$\epsilon_{\mathrm{x}} = \frac{du(x)}{dx} \tag{4.4}$$

From Eqs. (4.2)–(4.4), we can obtain the relationship between the applied force and the displacement of the bar as

$$P = YA\frac{du(x)}{dx} \tag{4.5}$$

or by using the applied distributed force per unit length along the bar, p_{x}, we have

$$p_{\mathrm{x}} = YA\frac{d^2u(x)}{dx^2} \tag{4.6}$$

We can easily understand from Eqs. (4.5) and (4.6) that the axial rigidity of the bar is defined by YA. This modeling technique is quite simple and basic. However, it is expedient and effective to model carbon–carbon (C–C) bond mechanics. Therefore, there have been many analytical research studies using bar model to computationally measure Young's modulus and to investigated the structural behaviors of nanotubes.

4.4 Bending Rigidity for Beams

In many cases for engineering, structural members must resist pure bending moments, and forces applied not only axially but also laterally or transversely to their axis. Such structural members are called "beams." With modern materials, the beam is a quite useful structural member in many different types of structural applications and a dominant member for engineering construction.

Beam deformations due to bending are determined from deformations taking place along the beam. These are based on the kinematic hypothesis that plane cross sections through a beam remain plane during bending. The equation for axial curvature κ is expressed in terms of first and second derivatives of bending deformation v as follows:

$$\kappa = \frac{1}{\rho} = -\frac{\dfrac{d^2v}{dx^2}}{\left[1+\left(\dfrac{dv}{dx}\right)^2\right]^{3/2}} \tag{4.7}$$

where ρ is a radius of curvature, the reciprocal of which gives the curvature, and x expresses the beam axis. In general, since the beam deformations v are very small, the corresponding slopes dv/dx are also very small. As a result, in the denominator of Eq. (4.7), the term $[1+(\frac{dv}{dx})^2]^{3/2}$ is approximately equal to 1 because the square of slope is a negligible quantity in comparison with unity and the curvature is given approximated by

$$\kappa = \frac{1}{\rho} \approx -\frac{d^2v}{dx^2} \tag{4.8}$$

Equation (4.8) is fundamental kinematic relation of a beam. On the other hand, the well-known basic moment–curvature relationship of an elastic beam is given by

$$\kappa = \frac{M(x)}{EI} \tag{4.9}$$

From Eqs. (4.8) and (4.8), bending deformation v of a homogeneous elastic beam is described by the Bernoulli–Euler equation [72]

$$YI\frac{d^2v}{dx^2} = -M(x) \tag{4.10}$$

with $M(x)$ the bending moment at x and I the area moment of inertia (or the second moment of area) with respect to the axis of bending, Eq. (4.10) is an approximation for small elastic bending deformation of beams and deflections due to shearing force are not considered in this equation. However, it is well known that this gives sufficiently accurate values for engineering applications provided that the beams are long compared with their cross-sectional dimensions and moment effects predominate as compared with the shear effect. It should be emphasized from Eq. (4.10) that the bending behavior of beams is determined by the bending rigidity YI, which is thus given by the product of Young's modulus, which is an inherent property of a material, and the area moment of inertia, which is strongly dependent on the beam cross-sectional shape and dimension. For a tube with outer diameter D_{out} and inner diameter D_{in}, for example, we have a value of

$$I = \frac{\pi}{64}\left(D_{\text{out}}^4 - D_{\text{in}}^4\right) \tag{4.11}$$

and for rectangular cross section whose base is b and whose height is h, the quantity is

$$I = \frac{bh}{12}. \tag{4.12}$$

In order to solve Eq. (4.10), boundary conditions at both end of the beam must be prescribed. There are three typical support conditions such as the following:

(i) roller and pinned support
(ii) clamped and fixed support
(iii) free end

In the case of (i), the displacement v and the bending moment M must vanish. Hence,

$$v = 0 \text{ and } \frac{d^2v}{dx^2} = 0.$$

In the case of (ii), not only v but also the slope dv/dx vanish because the beam perfectly fixed in this support. A support of (iii) is free of the bending moment and the shearing force. Therefore

$$\frac{d^2v}{dx^2} = 0 \text{ and } \frac{d^3v}{dx^3} = 0.$$

We can calculate the deflection v by direct integration of Eq. (4.10) and making use of the boundary conditions mentioned above. Let us introduce below some examples of bending deformations under typical loading and boundary conditions.

When a concentrated load P is applied at the free end of a cantilever beam (fixed at one end and free at the other end, shown in Section 2.4) of length L, we find the deflection v at free end δ

$$\delta = \frac{PL^3}{3YI}. \tag{4.13}$$

In the case of a cantilever beam that supports a uniformly distributed downward load q (see the right panel in Fig. 2.3), we obtain

$$\delta = \frac{qL^4}{8YI}. \tag{4.14}$$

For a beam fixed at both ends (see Fig. 2.4),

$$\delta = \frac{PL^3}{192YI}. \tag{4.15}$$

For a simply supported beam

$$\delta = \frac{PL^3}{48YI}. \tag{4.16}$$

We can easily find from Eq. (4.13)–(4.16) that the deformation of the beam drastically increases as beam length L and strongly depends on the cross-sectional shape and dimension. The beam theory has thus been frequently used to model not only C–C bonds but also the carbon nanotube itself.

4.5 Axial and Bending Rigidity for Shells

Structural modeling as thin-walled curved and cylindrical shell components is also widely used in many fields, such as civil, architectural, offshore, and mechanical engineering and aircraft structures. The examples include pipelines, pressure vessels, airplane, and a wide variety of spatial structures.

From the structural mechanics point of view, trusses and beams shown in the preceding two sections are considered one-dimensional members. However, to regard a nanotube as a beam is inadequate, especially when the deflection is localized or restricted. The application of the cylindrical shell theory to nanotube modeling is thus reasonable. In this section, the bending theory of the shell is briefly introduced (the formulation processes are not described here; or the details, see [86]).

In basic and classical Kirchhoff hypothesis, deformation behaviors of cylindrical shells subjected to external loads are dependent on two quantities, tension and bending rigidities, which can be determined by Young's modulus, Poisson's ratio, and wall thickness. The important point to remember is that Y and h are uniquely determined by two non-equivalent equations describing axial and bending deformations. Neglecting curvature, i.e., supposing the tube to be equivalent to a sheet of width $2\pi R$, Y and h can be deduced from the bending rigidity D and in-plane stiffness C of a graphene plate [72]

$$D = \frac{Yh^3}{12(1-\nu^2)}, \tag{4.17}$$

and

$$C = Yh. \tag{4.18}$$

4.6 Young's Modulus Prediction

4.6.1 *C–C Bond Modeling Using Truss and Beam Elements*

In the following, we review some research studies conducted to predict Young's modulus of nanotubes based on continuum-based elastic theories.

The model introduced here is based on the assumption that the bonds between carbon atoms (C–C bond) are connected by load-carrying members and behave like space-frame structures when subjected to external loading. The modeling images are illustrated in Fig. 4.1. The linkage technique was initially proposed by Odegard *et al.* [74]. In this article, the nearest-neighboring C–C bond is modeled as equivalent continuum pin-jointed truss elements, which represent the forces between two atoms. In truss elements, only longitudinal tension and compression forces and corresponding displacements in the axial direction along the truss can be

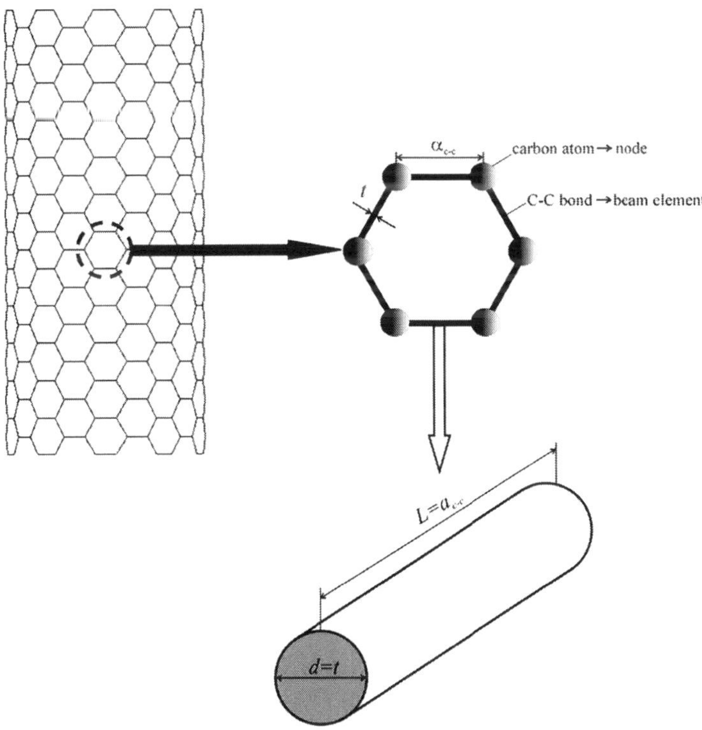

Figure 4.1 Simulation of an SWNT as a space-frame structure. In this technique, carbon atoms C–C bonds are modeled as nodes and beam elements, respectively. Reprinted from *Composites B Eng.*, **36**(5), Tserpes, K. I., and Papanikos, P., Finite element modeling of single-walled carbon nanotubes, 468–477, Copyright (2005), with permission from Elsevier.

considered. Atoms in a lattice have been viewed as masses that are held in place with atomic forces that resembles elastic springs. The method provides a link between computational chemistry to predict the molecular properties of nanostructured materials and solid mechanics describing the macroscopic mechanical properties of materials.

A developed concept in which the C–C bonds between two nearest-neighboring atoms act like load-bearing beam members and an individual atom is modeled as the joint of the related beam members was successfully proposed by Li and Chou [82]. Contrary to the truss elements, not only axial but also bending and torsion rigidities can be modeled in beam elements.

Based on the concept of the above study by Li and Chou [82], a three-dimensional finite element model for armchair, zigzag, and chiral SWNTs is proposed by Tserpes and Papanikos [87]. In this study, ANSYS commercial finite element code [88] was used to develop the three-dimensional beam model. Moreover, the non-linear spring (truss) model has also been introduced by Kalamkarov *et al.* [89]. In their study, the numerical prediction of the mechanical properties of not only SWNTs but also MWNTs has been successfully developed. These studies thus suggest that finite element approaches can be a valuable tool to computationally investigate the mechanical properties of nanotubes.

In the above-mentioned truss models developed by Odegard *et al.* [74], beam models by Li and Chou [82], and similar approaches, however, the bending stiffness of the graphite layer in nanotubes was not taken into consideration. For this reason, the applications of their models have been restricted to simulations for the uniform elongation to analytically predict the elastic modulus of nanotubes and in which the local bending deformation is quite insignificant. In recent years, as a modified structural mechanics model using the beam element, the approach that includes the bond inversion energy has been developed by Wan and Delale [90]. See Fig. 4.2 for the schematic illustration of the atomic interactions considered in the model.

4.6.2 *Modeling Using Beams and Shells*

In the case that local deformations of nanotubes should not be considered (e.g., Euler buckling behaviors under axial compression [91]), it is reasonable to model the nanotubes themselves as beam members because of their dimensions (lengths are extremely long compared with its diameter). Moreover, if cross-sectional deformations or local collapses on nanotubes are induced by external loads, the above-mentioned beam theory is not appropriate to apply anymore and the shell theory must be alternative and realistic options. Fig. 4.3 shows the linkage modeling and the equivalence of molecular, finite element, and continuum models.

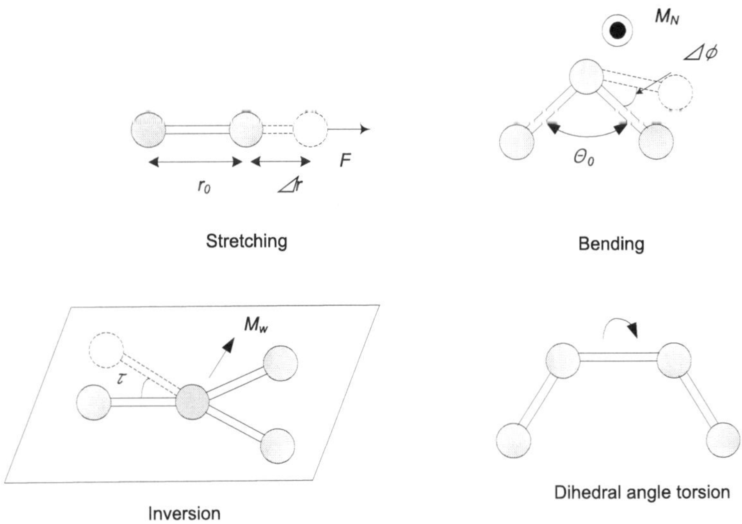

Figure 4.2 Inter-atomic interactions in molecular mechanics. Deformations such as stretching, bending, and torsion can be modeled by this approach, which includes the bond inversion. Adapted from Ref. [90].

In this figure, the finite element modeling corresponds to the C–C bond modeling using trusses or beams shown in Section 4.6.1. Here we consider the continuum-based shell model in Fig. 4.3. Analytical approaches to investigate the buckling behaviors of nanotubes using the beam model are described in Chapter 5.

Concerning an SWNT, the main question is whether it can be described as a homogeneous tube made from an isotropic shell of finite thickness h (to be determined), whose deformation can be calculated using classical equations and two independent elastic parameters (e.g., Y and ν), without requiring atomistic calculations at least in the linear regime.

Many research studies have been conducted on the modeling of nanotubes as cylindrical shells (and graphenes as continuous plates) and the evaluation of their "effective thickness" thus far. However, the reported Young's moduli and thicknesses in a large number of research studies are extremely scattered, as shown in Table 4.1.

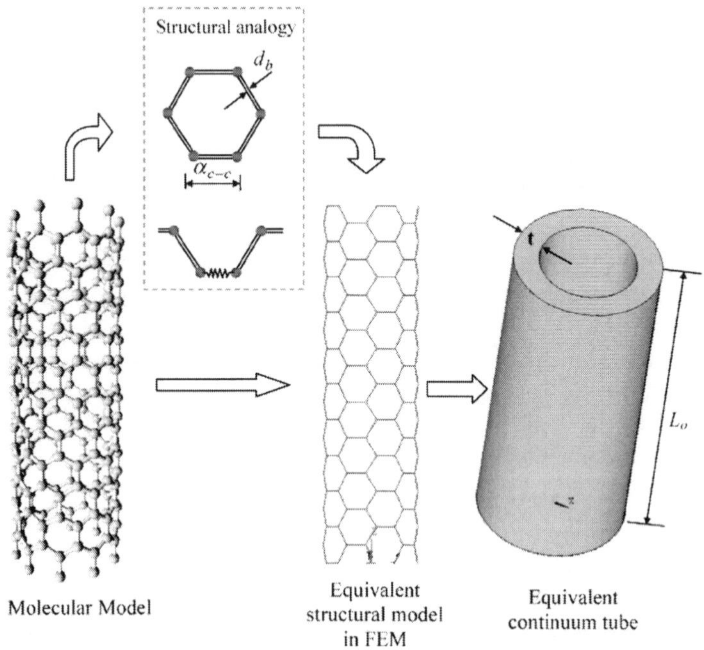

Figure 4.3 Equivalence of molecular, finite element and continuum models. Reprinted from *Int. J. Solids Struct.*, **43**(22–23), Kalamkarov, A. L., Georgiades, A. V., Rokkam, S. K., Veedu, V. P., and Ghasemi-Nejhad, M. N., Analytical and numerical techniques to predict carbon nanotubes properties, 6832–6854, Copyright (2006), with permission from Elsevier.

Answering this question is thus not straightforward, but a positive consensus emerges after a careful examination of various calculations and experiments performed thus far.

As a first order approximation, therefore, we treat a nanotube as a homogeneous, isotropic, and continuous shell of thickness h, so that all its elastic deformations can be calculated knowing Young's modulus Y and Poisson's ratio ν.

D, C, and ν are intrinsic parameters of graphene essentially determined by C–C bond rigidity, that can be calculated without any assumption on h.

The bending rigidity of a graphene sheet D has been approximated using classical elasticity and the out-of-plane vibration frequency of graphite as measured by neutron scattering [92]. It

Table 4.1 The thickness and Young's modulus of carbon nanotubes.

Authors	Method	Wall thickness (nm)	Young's modulus (TPa)
Lu (1997) [61]	Molecular dynamics	0.34	0.974
Hernández *et al.* (1998) [63]	Tight binding molecular dynamics	0.34	1.24
Odegard *et al.* (2002) [74]	Equivalent-continuum modeling	0.69	
Li and Chou (2003) [82]	Structural mechanics (stiffness matrix method)	0.34	1.01
Jin and Yuan [75]	Molecular dynamics	0.34	1.238
Tserpes and Papanikos (2005) [87]	Structural mechanics (finite element method)	0.147	
Yakobson *et al.* (1996) [32]	Molecular dynamics	0.066	5.5
Zhou *et al.* (2000) [97]	Tight-binding model	0.074	5.1
Kudin *et al.* (2001) [43]	*Ab initio* computations	0.089	3.859
Tu and Ou-Yang (2002) [93]	Local density approximation model	0.075	4.7
Vodenitcharova and Zhang (2003) [76]	Ring theory continuum mechanics	0.0617	4.88
Pantano *et al.* (2004) [118]	Continuum shell modeling	0.075	4.88
Goupalov (2005) [77]	Continuum model for wavelength phonons	0.087	
Wang *et al.* (2005) [78]	*Ab initio* calculation	0.0665	5.07

Source: Data from Ref. [94].

has also been estimated via calculation of the strain energy using a force-field framework, i.e., a molecular mechanics approach [74], molecular dynamics, and first-principle calculations [35, 59].[a] D is found to vary between 0.85 and 1.5 eV depending on the method used. In-plane (or axial) stiffness C is usually derived from the

[a]The important point is that bending rigidity of a graphene sheet is determined by inversion configuration of the C–C bonds [74]. Without this term, the bending rigidity of the graphene sheet should be zero.

second derivative of the strain energy E_s with respect to the axial strain ε [33, 35, 59, 69],

$$C = \frac{d^2 E_s}{d\varepsilon^2}.$$

C is found to be close to 60 eV/atom and seems less dependent on the calculation method than D. Poisson's ratio ν has been also calculated in a number of papers [33, 35, 69, 93] with a dispersion of values, between 0.16 and 0.34.

Combining Eqs. (4.17) and (4.18), one finds

$$Y \simeq 5 \text{ TPa} \quad \text{and} \quad h \simeq 0.07 \text{ nm}.$$

There is thus no doubt that the effective thickness of an SWNT in an isotropic continuous shell approximation, is less than 0.1 nm, similar to that of a single graphene layer. Many authors use, however, $h = 0.34$ nm and $Y = 1$ TPa, since it compares directly with bulk graphene values.

4.6.3 *Bridging Atomistic and Continuum Modeling*

In addition to the discussion in the previous subsection, it is quite effective if the relationship and analogy between atomistic and continuum modelings can be considered analytically. To do this, analytical quantities of tension and bending rigidities of SWNTs must be obtained in terms of parameters in the empirical interatomic potential such as the Tersoff–Brenner potential introduced in the previous chapter. Accordingly, elastic bending rigidity D of nanotubes has been obtained by Huang *et al.* [94] as

$$D = \frac{\sqrt{3}}{2} \left. \frac{\partial V_{\mathrm{ij}}(r_{\mathrm{ij}}; \cos\theta_{\mathrm{ijk}}, k \neq i, j)}{\partial \cos\theta_{\mathrm{ijk}}} \right|_0 \qquad (4.19)$$

for a pair of carbon atoms i and j, where V_{ij} is the interatomic potential functions, r_{ij} is the bond length, θ_{ijk} is the bond angle between i–j and i–k bonds, k represents the atoms in the vicinity, and the subscript 0 denotes the unstrained equilibrium state. It should be noted that Young's modulus and the thickness are dependent on the selected empirical interatomic potential, loading conditions (e.g., uniaxial tension, uniaxial stretching, equibiaxial stretching), the nanotube radius, and its chirality when the tube radius is less than 1 nm. Huang *et al.*'s [94] article concludes that this is the reason why Young's moduli and thicknesses reported by many researchers shown in Table 4.1 are so scattered.

Another important point that Huang *et al.* [94] showed is that it is unnecessary to define the SWNT Young's modulus and thickness separately since all experimentally measurable or theoretically calculable properties involve Yh, not E nor h separately. In fact, the prior modeling and simulations of SWNT can be grouped to two types: One takes the interlayer spacing of graphite 0.34 nm as the SWNT thickness, and the resulting effective Young's modulus is around 1 TPa [33, 61, 82]; the other is based on the continuum shell modeling, which gives the effective thickness around 0.066 nm and Young's modulus around 5.5 TPa as discussed above [32, 43, 95–97]. These two types, however, give approximately the same product Yh, about 3.4 TPa·nm.

Recently, the developed analytical formula has been also derived for the elastic bending rigidity D of monolayer graphene by considering multibody atomistic interactions up to the third nearest neighbors [98]. This can be achieved by accounting for in a bond-order empirical interatomic potential by the effect of not only the bond angles (second nearest neighbors) but also the dihedral angles (third nearest neighbors).

4.7 Cauchy–Born Rule

It has to be well understood that this continuum approach remains approximative, since defining a thickness for one-atom-thick layers does not have a rigorous physical meaning, even when using the notion of an electronic density cloud. It should not, however, be concluded that a continuum elastic model cannot be applied rigorously. Arroyo and Belytshko [80] proposed a method to describe one-atom-thick layer deformation extending hyper-elastic models used for bulk materials. They also show the limitations of continuum models, especially in the non-linear elastic regime.

The main disadvantage of the shell model is that it has neglected the atomic characteristics. The effect of chirality is only modeled through different radii, which is insufficient in many cases since nanotubes with similar radii but different chiralities (such as those in Fig. 4.1) may exhibit distinct behaviors [99]. Moreover, it cannot be used to study the effect of defects on the mechanical properties, and is unsuitable to account for the forces acting on individual atoms [100]. To incorporate the effect of chirality, Zhang *et al.* [101, 102] have developed a continuum theory by directly incorporating the interatomic potentials into the constitutive model for nanotubes. This continuum theory agrees well with the atomistic simulations, such as the fracture strain of nanotubes [103–105] and nanotube reinforced composites [106], defect nucleation in nanotubes [107] and boron-nitride nanotubes [108]. Such an atomistic-based continuum theory can account for the finite temperature [109, 110] as well as the coupled electromechanical behavior [111, 112].

The Cauchy–Born rule is effective at linking the deformation of an atomic system to that of a continuum without other phenomenological input. Here the Cauchy–Born rule is outlined briefly (see also [113] for the detailed description).

Figure 4.4 illustrates an image of the direct map between undeformed and deformed atoms. As can be found in Fig. 4.4, in this rule the exact expression for the deformed bond vector can be related to the undeformed one by the appropriate mapping. This assumption successfully describes crystal behaviors in the case that the continuum deformation is nearly homogeneous in the scale of the crystal. Thus, the bond vector for pair atoms is written according to the local deformation gradient as

$$\mathbf{a} = \mathbf{F}(\mathbf{X}) \cdot \mathbf{A}, \tag{4.20}$$

where $\mathbf{A}$ and $\mathbf{a}$ are the undeformed and deformed bond vector, $\mathbf{X}$ is the fixed coordinate and $\mathbf{F}$ denotes the local deformation gradient at $\mathbf{X}$. By using the map $\mathcal{F}$, we can obtain the following expression:

$$\mathbf{a} = \mathcal{F}(\mathbf{X} + \mathbf{A}) - \mathcal{F}(\mathbf{X}). \tag{4.21}$$

By expanding the right-hand side of Eq. (4.21) by the Tailor series with regard to $\mathbf{X}$ and using the relation $\mathbf{F} = \nabla\mathcal{F}$, we have

$$\mathbf{a} = \mathbf{F}(\mathbf{X}) \cdot \mathbf{A} + \frac{1}{2!}\nabla\mathbf{F}(\mathbf{X}) : (\mathbf{A} \otimes \mathbf{A}) + \frac{1}{3!}\nabla\nabla\mathbf{F}(\mathbf{X})\lozenge(\mathbf{A} \otimes \mathbf{A} \otimes \mathbf{A})$$
$$+(\text{higher-order terms}). \tag{4.22}$$

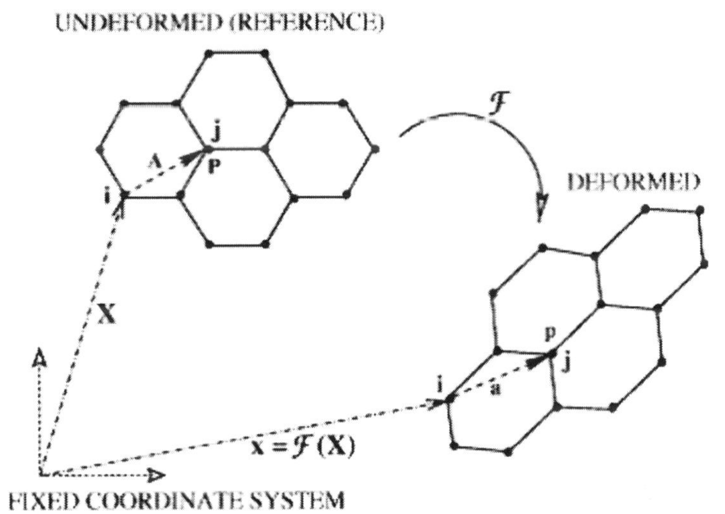

Figure 4.4 The direct map. Reprinted from *Comput. Mater. Sci.*, **40**(1), Chandraseker, K., and Mukherjee, S., Atomistic-continuum and ab initio estimation of the elastic moduli of single-walled carbon nanotubes, 147–158, Copyright (2007), with permission from Elsevier.

in which the notation ":" is the action of a third rank tensor on a second rank tensor that results in a vector, "◊" corresponds to the action of a fourth rank tensor on a third rank tensor that results in a vector, and "⊗" denotes the standard tensor product. This equation gives the exact expression for the deformed bond vector. If the deformation gradient is constant in space, all its gradients vanish, and the Taylor series reduces to the standard Cauchy–Borne rule, which becomes exact in this case.

Thus, we can see that the Cauchy–Born rule does not neglect the atomic characteristics and therefore can simultaneously retain both its computational efficiency and correctness. Under the fundamental assumption above, the Cauchy–Born rule has been successfully applied to link the atomistic and continuum modeling of SWNTs [99, 104, 114]. In a complementary continuum study, Arroyo and Belytschko developed continuum theories of nanotubes based on the Cauchy–Born rule [71, 115]. Moreover, Chandraseker *et al.* present the modified Cauchy–Born rule to obtain the deformation of SWNTs, which are assumed to be inhomogeneous at the atomic scale [113].

4.8 Multishell Nanotube Elasticity

We first consider a two-layer graphene slab as an anisotropic shell of thickness $h_{\mathrm{D}} = h + d_{\mathrm{i}}$, where $d_{\mathrm{i}} \simeq 0.34$ nm is the interlayer distance. Supposing that no sliding can occur between the two layers thanks to van der Waal forces, the neutral plane is located between the layers and the bending stiffness is determined now by the in-plane stiffness of C–C bonds [116], unlike the bending stiffness of a single graphene sheet that depends essentially on the out-of-plane stiffness. The two-layer slab (or tube) is well-described by a shell of Young's modulus $\simeq 1$ TPa and thickness about 0.34 nm.

More precisely, it was found that Young's modulus shows a discontinuous behavior as a function of the number of shells N [93]:

$$Y_{\mathrm{N}} = \frac{Nh}{(N-1)d_{\mathrm{i}} + h} Y$$

with $Y = 4.7$ TPa and $h = 0.074$ nm.

To assemble SWNTs into a MWNT, the non-bonded interaction between neighboring tube layers can be modeled as a Lennard–Jones type of interface potential, which introduces repulsive or attractive pressure along the shell surface as the system is deformed [117, 118]. A similar approach can also be readily extended to nanotube bundles.

4.9 Crystal of SWNTs

Bulk SWNT materials are usually made of ropes and bundles, not individual nanotubes. While covalent C–C bonds ensure strong intratube stiffness, transverse properties are governed by van der Waals (vdW) forces very similar to those at work in graphite interplane cohesion. The SWNT crystal is thus highly anisotropic, with low torsion and shear moduli. Elastic constants can be approximated with classical elasticity theory using those of graphite, in good agreement with more rigorous calculations [119, 120]. Supposing in-plane stiffness is the same for graphite and for SWNT, it is straightforward that

$$\frac{Y_{\text{bundle}}}{Y_{\text{graphite}}} = \frac{\pi D d_{\text{i}}}{S},$$

where D is the SWNT diameter, d_{i} is the interplane spacing in graphite, and S is the area of the unit cell in bundles, i.e.,

$$S = (D + d_{\text{i}})^2 \cos\left(\frac{\pi}{6}\right).$$

So we get Young's modulus close to 650 GPa for a 1.4 nm diameter SWNT bundle. We can estimate as well a shear modulus of 20 GPa.

It is thus found that the various forms of carbon nanotubular structures exhibit different Young's modulus according to their structure. This should be taken into account when determining the elastic properties of any nanotube-based nanocomposites.

Chapter 5

Buckling

Age imprints more wrinkles in the mind than it does on the face.
— Michel de Montaigne, in "Essais"

5.1 Better Bend than Break

The previous three chapters focus mainly on the linear elasticity of carbon nanotubes at small deformation. Differently, the non-linear mechanical response of nanotubes under large deformation has also long received wide attention. In particular, their buckling behaviors have been regarded as a major topic in the nanotube research community [121, 122].

In terminology, buckling means a deformation process in which a structure subjected to high stress undergoes a sudden failure, showing visibly large displacements transverse to the load [123]. A typical example of buckling may be observed when pressing the opposite edges of a thin long elastic beam toward one another (see Fig. 5.1). For small loads, it is compressed in the axial direction with keeping the linear shape, in which the strain energy is proportional to the square of the axial displacement. Beyond a certain critical load, however, it suddenly bends arch-wise and

Elastic and Plastic Deformation of Carbon Nanotubes
Hiroyuki Shima and Motohiro Sato

ISBN 978-981-4310-75-8 (Hardcover), 978-981-4364-15-7 (eBook)
www.panstanford.com

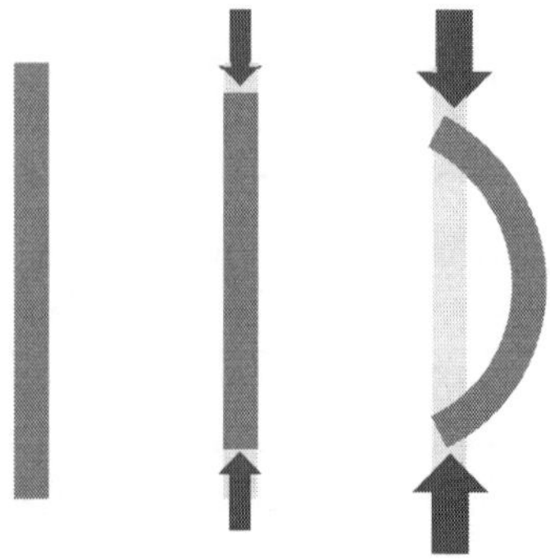

Figure 5.1 Schematic diagram of buckling of an elastic beam under axial compression. (a) Pristine beam; (b) Axial compression for small load; (c) Axial buckling observed beyond a critical load.

the relation between the strain energy and displacements deviates significantly from the square law mentioned above.[a] In addition to axial compression, bending and torsion also can give rise to buckling behaviors of elastic beams, where the deformation patterns obtained after buckling strongly depend on the geometric and material parameters.

The above-mentioned argument on macroscopic elastic beams encourages us to explore what buckled patterns are obtained in carbon nanotubes as well as the similarities and differences compared with macroscopic counterparts. Thus far, tremendous efforts have been devoted to the buckling analysis of carbon nanotubes under diverse loading conditions: compression [32, 124–133], bending [11, 31, 134–140], torsion [141–145], and their certain combinations [146–150]. Such extensive studies have been driven by (i) the exceptional resilience of nanotubes against buckling (i.e., their shapes are reversible upon unloading) and (ii) the substantial influence of buckling on their mechanical and physical properties. Still, the experimental investigation of nanotube

[a]The buckling mentioned here is similar to pure bending in the sense that both of them involve bending moments. In bending, the moments are essentially independent of the resulting deflections. In the buckling, in contrast, the moments and deflections are mutually dependent so that neither the moments nor deflections are proportional to loads.

buckling remains a challenge because of difficulties in manipulation at the nanoscale [11]. For that matter, both theoretical and numerical approaches have been used to explore the buckling behavior of nanotubes. With the background in mind, we shall review in this section research developments on the nanotube buckling during the past two decades.

5.2 Resilience and Sensitivity

To date, many intriguing post-buckling morphologies of nanotubes have been widely predicted in theory and observed in experiments: local kinks in bent SWNTs [32], belt-like structures in strongly twisted SWNTs [32], and the rippling or Yoshimura (diamond-shaped) patterns in tens-walled nanotubes under various loading conditions [11, 96, 151–159]. One of the most outstanding features of post-buckling nanotubes is their remarkable buckling capacity. Indeed, experimental investigations have shown that the buckling deformation can be completely recovered when the load is removed [11, 31, 134, 151, 160–162]. The marked resilience is primary because of (i) the large in-plane rigidity of graphene sheets rather than relatively low bending rigidity [163] and (ii) the intrinsic hollow geometry with extremely large aspect ratio that carbon nanotubes exhibit.

Apart from the resilience, physical and mechanical properties of carbon nanotube are susceptible to buckling; namely, the breakdown of the structural symmetry due to the buckling may trigger sudden changes in physical properties. Only a few to mention are the thermal conductivity reduction [164] and the radial breathing mode frequency shift [165] under axial strain, an emergence of interwall sp^3 bondings under bending [166], and electromechanical responses under bending [167–169] and torsion [202]. Besides, the buckling-induced reduction in nanotubes stiffness will not only impair their ability to sustain external loadings as reinforced fibers in nanocomposites [151, 161] but also give rise to large uncertainties in the vibration behavior of nanotubes as nanoscale resonators [11, 170]. These buckling-property relations can significantly influence the performance of nanotubes as structural or functional elements, thus being the primary reason why so much effort has been made on the study of nanotube buckling.

5.3 Bend Buckling of SWNTs

5.3.1 *Kink Formation*

The buckling of SWNTs under bending was pioneered in 1996 [31] using high resolution electron microscopes and molecular-dynamics (MD) simulation. The overall scenario is summarized as follows. By bending an initially straight SWNT, its outer and inner sides undergo stretching and compression, respectively. Prior to buckling, the strain energy increases as a quadratic function of the bending angle, corresponding to harmonic deformation. When the bending angle exceeds a threshold, the nanotube loses its structural stability via the formation of kinks at the compressed inner side where the bending curvature is largest.[a] Following the formation of the kink, the strain energy increases approximately linearly until bond breaking occurs under quite large deformation.[b]

Figure 5.2(a) shows a transmission electron microscope (TEM) image of a bent SWNT of diameter 1.2 nm [31]. It has developed a single kink in the middle, through which the strain energy on the compressed sides is released. Upon removing the bending moment at this bending angle of 120°, it returns to the initial structure form completely without any permanent damage such as bond breaking and defects. This observation clearly proves that SWNTs possess extraordinary structural elastic flexibility.

Figure 5.2(b) presents computer-simulated reproduction of the kink experimentally observed, providing atomistic and energetic information about the bending process. The overall shape of the kink, along with the distance of the tip of the kink from the upper wall of the tube, is in quantitative agreement with the TEM picture

[a] After buckling, more kinks appear with increasing curvature as demonstrated in Ref. [117].

[b] A similar characteristic energy-strain curve, an initial quadratic curve followed by a linear increase, arises in the case of axial compression as demonstrated in Ref. [32].

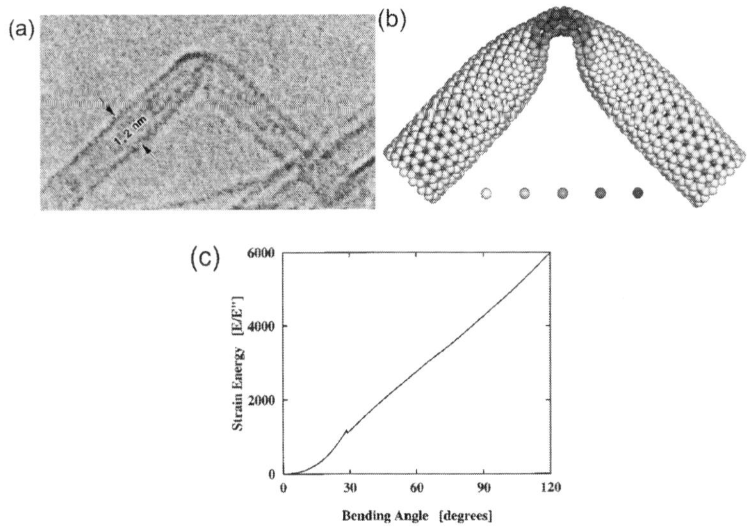

Figure 5.2 (a) Kink structure formed in an SWNT with diameters of 1.2 nm under bending. The gap between the tip of the kink and the upper wall is about 0.4 nm. (b) Atomic structure of a single kink reproduced by computer simulations. The shaded circles beneath the tube image express the local strain energy at the various atoms, measured relative to a relaxed atom in an infinite graphene sheet. The strain energy scale ranges from 0 to 1.2 eV/atom, from left to right. (c) The total strain energy (in dimensionless units) of an SWNT of diameter ~1.2 nm as a function of the bending angle up to 120°. The dip at ~30° in the curve is associated with the formation of a kink. Reprinted with permission from Iijima, S., Brabec, C., Maiti, A., and Bernholc, J., *et al.* (1996) Structural flexibility of carbon nanotubes, *J. Chem. Phys.*, 104(5), 2089–2092. Copyright 1996, American Institute of Physics.

of Fig. 5.2(a).[a] The coding denotes the local strain energy at the various atoms, measured relative to a relaxed atom in an infinite graphene sheet. In all simulations, the same generic features appear: For bending below a sharp critical curvature (~30°), the tube obeys Hooke's law, and the energy increases quadratically as a function of the bending angle (see Fig. 5.2(c)). In this harmonic regime, the hexagonal rings on the tube surface are only slightly strained with maintaining the hexagonal carbon network. Beyond the critical

[a]This latter distance is about 0.35 nm, precisely where the van der Waals interaction becomes strongly repulsive.

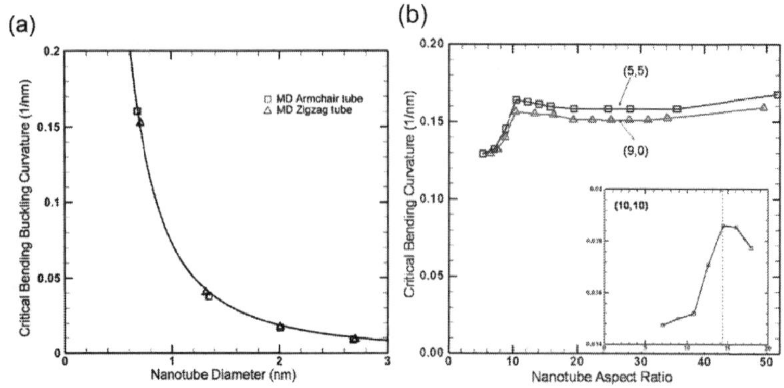

Figure 5.3 (a) Relationship between critical bending buckling curvature and nanotube diameter obtained from MD analyses. The tube length is fixed at 24 nm. (b) The length dependence of critical bending buckling curvature of (5,5), (9,0) and (10,10) tubes. Reprinted with permission from Cao, G., and Chen, X., *Phys. Rev. B*, **73**, 155435-1–10, 2006. Copyright (2006) by the American Physical Society.

curvature, the excess strain on the compressed side reaches a maximum and is released through the formation of a kink that increases the surface area of the bending side. This is accompanied by a dip in the energy versus bending angle curve, as shown in Fig. 5.2(c).

5.3.2 *Diameter Dependence*

The geometrical size has significant effects on the buckling behaviors of SWNTs under bending. For instance, those with a small diameter can always sustain a larger bending angle prior to buckling [117]. It was revealed by MD simulations that the critical bending buckling curvature κ_c monotonically increases with reducing the nanotube diameter d as displayed in Fig. 5.3(a). The relationship between κ_c and d can be fitted as [32, 117] $\kappa_c \propto d^{-2}$, which holds for no matter what chirality the nanotube has.

In addition to the diameter dependence, κ_c of SWNTs is affected by the aspect ratio, defined by the length$(=L)$/diameter$(=d)$. It follows from Fig. 5.3(b) that κ_c increases with L/d followed by saturating a constant in the limit of infinitely long tube [117].

The results show that κ_c is almost constant for sufficiently long nanotubes such that $10 < L/d < 50$ or more, while it drops off for short nanotubes satisfying $L/d < 10$. The critical aspect ratio that separate the two regions is sensitive to the tube diameter (as implied by the inset of Fig. 5.3(b)), but it is almost independent of chirality.

5.3.3 *Transient Bending*

We have learned in Section 5.3.1 that for relatively thin SWNTs, the buckling is characterized by a change in the functional dependence of the strain energy on bending angle. The dependence was quadratic in the prior-buckling regime and linear in the post-buckling regime, in which the buckling point was marked by the appearance of a discontinuity in the potential energy curve [31, 32]. However, this is not the case for larger diameter nanotubes. As the diameter is increased, a second discontinuity appears in the potential energy curve at a larger bending angle than the first one. The origin of the two discontinuities is accounted for by Fig. 5.4. The bottom image in the figure shows a just-buckled wall of a (30, 30) SWNT corresponding to the first discontinuity in the potential curve, in which the buckled side is far from the opposite side. Therefore, more bending is required to bring the two sides close enough (i.e., to the equilibrium van der Waals (vdW) distance), which results in the kink formation associated with the second discontinuity.

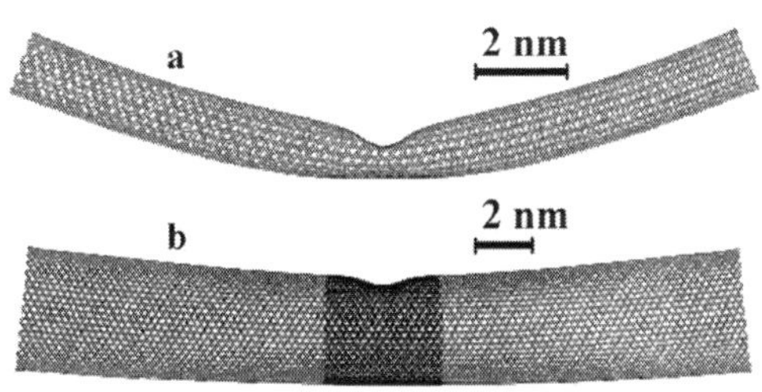

Figure 5.4 Predicted shape of SWNTs just after buckling, based on MD simulations. (a) 15.7 nm-long (10, 10) SWNT at the bending angle $\theta = 43°$ and (b) 23.6 nm-long (30, 30) SWNT at $\theta = 23°$. Note the difference in scale. Reprinted with permission from Kutana, A., and Giapis, K. P., *Phys. Rev. Lett.*, **97**, 245501-1–4, 2006. Copyright (2006) by the American Physical Society.

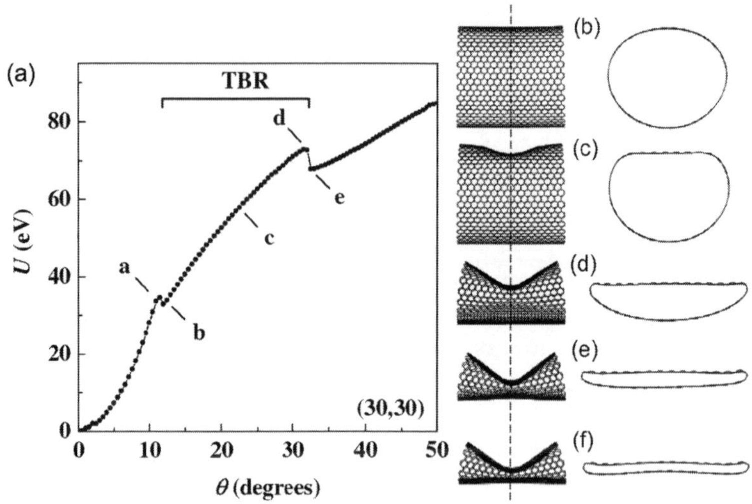

Figure 5.5 (a) Strain energy U for a 23.6 nm-long (30,30) SWNT as a function of the bending angle θ. The letters a–e indicate the points for which the tube shape and cross section at the buckling point are shown in the right panel. "TBR" denotes the transient bending regime. (b)–(f) Side views and buckling point cross sections of a 23.6 nm-long (30,30) SWNT at various bending angles corresponding to the points a–e marked on the strain energy curve in the plot (a). Only the portion near the buckling region is shown for the longitudinal views. Bending occurs in the plane of the page. Reprinted with permission from Kutana, A., and Giapis, K. P., *Phys. Rev. Lett.*, **97**, 245501-1–4, 2006. Copyright (2006) by the American Physical Society.

The above-mentioned buckling behavior is illustrated in Fig. 5.5(a), where the strain energy U is plotted as a function of bending angle θ for a (30, 30) SWNT. Three distinct deformation regimes are observed, clearly separated by discontinuity points at $\theta = 12^\circ$ and 32°. In the initial elastic regime, U exhibits a quadratic dependence on θ, while the cross section experiences progressive ovalization as the bending angle increases, culminating to the shape in Fig. 5.5(b). The buckling event is marked by an abrupt transition from the oval cross section to one with the flat top shown in Fig. 5.5(c). As the bending angle increases beyond the first discontinuity (i.e., during the transient bending regime (TBR) indicated in Fig. 5.5(a)), the flat portion of the top wall expands continuously across the nanotube (Figs. 5.5(d) and (e)).[a] As a result, the top-to-bottom wall distance decreases gradually, reducing the tube cross section at the buckling site. The strain energy curve in TBR is no longer quadratic; in fact, the exponent becomes less than unity. When the approaching opposite walls reach the vdW equilibrium distance of 3.4 Å (Fig. 5.5(f)), the cross section collapses, forming the kink, and a second discontinuity is observed.

5.4 Bend Buckling of MWNTs

5.4.1 *Ripples Emerging*

Following the discussion on SWNTs, we look in this section into the non-linear mechanics of MWNTs under bending. The difference between the mechanical responses of SWNTs and MWNTs lies in the presence of vdW interactions between the constituent carbon layers. Thus, the effect of the multi-layer structure on MWNTs' buckling behavior has been a subject of great research interest. Apparently, thicker MWNTs with tens of concentric walls may be stiffer than few-walled thin MWNTs against bending, since the inner tubes of MWNTs may reinforce the outer tubes via the normal vdW interaction. However, it is just the contrary. In fact, while MWNTs with small diameter exhibited a bending stiffness of around 1 TPa, those with larger diameter were much more compliant with the stiffness around 0.1 TPa [11]. This dramatic reduction in the bending stiffness was attributed to the so-called rippling effect, i.e., the emergence of the wavelike distortion on the inner arc of the bent nanotube [22, 23, 96, 151, 161].

Figure 5.6 presents a clear example of the rippled MWNT structure [161]. An enhanced image at the ripple region, given in the right panel, displays the fact that under high bending, nanotubes collapse to form kinks on the internal (compression) side of the bend. The amplitude of the ripple increased gradually from inner to

[a] Interestingly, the transient bending regime is fully reversible: If bending is stopped before the second discontinuity occurs, unbending recovers the cross-sectional shapes at the buckling point.

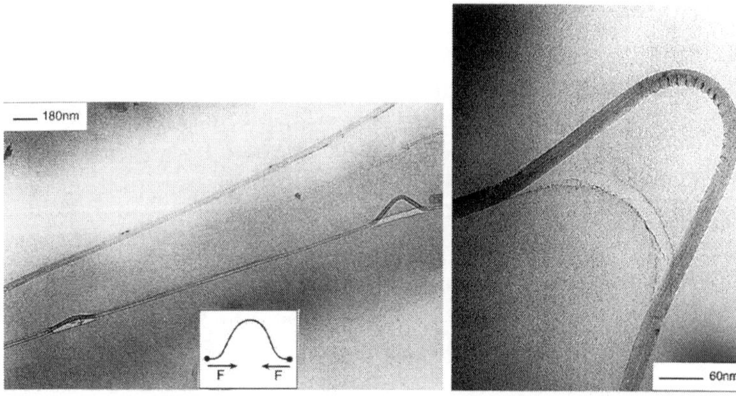

Figure 5.6 (Right) Micrographs of long and slender MWNTs under compression, showing that the nanotubes behave as elastic rods and form bends. (Left) Kinks appearing on the internal (compression) side of the bend MWNTs. Reprinted with permission from Lourie, O., Cox, D. M., and Wagner, H. D., *Phys. Rev. Lett.*, **81**, 1638–1641, 1998. Copyright (1998) by the American Physical Society.

outer walls, essentially 0 for the innermost core tube to about 2 to 3 nm for the outermost wall. Such rippling deformation induces a significant reduction in the bending modulus, as has been explained theoretically by solving the non-linear differential equations [171].

5.4.2 *Yoshimura Pattern*

Precise information about the membrane profile and the energetics of the rippling deformation, which are unavailable in experiments, can be extracted from the large-scale computer simulations. Figure 5.7(a) shows a longitudinal cross section of the equilibrium configuration. This image is the computational analog of the TEM slices of rippled thick nanotubes reported in the literature [11, 23]. The simulations reproduce very well the general features of the observed rippled nanotubes: Nearly periodic wavelike distortions, whose amplitudes vanish for the inner tubes and smoothly increase toward the outer layer.

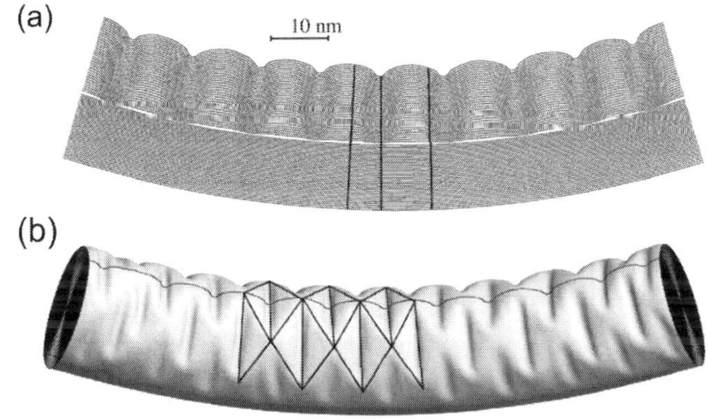

Figure 5.7 Rippling of a 34-walled carbon nanotube: (a) longitudinal section of the central part of the simulated nanotube. (b) The morphology of the rippled MWNT reminiscent of the Yoshimura pattern. The ridges and furrows have been highlighted, as well as the trace of the longitudinal section. Reprinted with permission from Arroyo, M., and Belytschko, T., *Phys. Rev. Lett.*, **91**, 215505-1–4, 2003. Copyright (2003) by the American Physical Society.

A remarkable finding in the simulations is that the rippling deformation closely resembles the Yoshimura pattern[a] (or called diamond buckling pattern). We can see that the rippling profile in Fig. 5.7(b) consists not of a simple linear sequence of kinks, but of a diamond-like configuration of kinks on the compressed side. Such Yoshimura pattern is well known as what characterizes the post-buckling behavior of cylindrical elastic shells at conventional macroscopic scale. The pattern has the interesting geometric property of being a nearly isometric mapping of the undeformed surface, at the expense of creating sharp ridges and furrows.

The rippling deformation, peculiar to thick MWNTs, is a consequence of an interplay between the relaxation of the strain energy through the geometric instability and the vdW energy that keeps the ripples in place. As intuitively understood, the bending flexibility of individual walls relative to the in-plane stiffness is very

[a]Yoshimura pattern is a special kind of surface deformation modes occurring in thin-walled cylindrical shells subjected to large lateral load. It is named after Yoshimura, a Japanese theoretician of the fracture mechanics, and its profile is characterized by a periodic diamond-like corrugation.

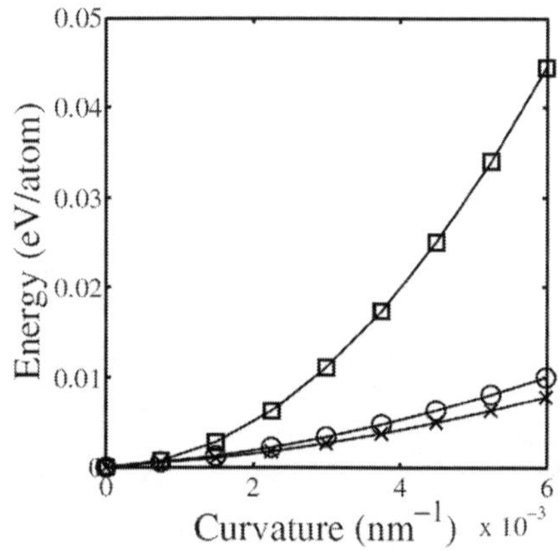

Figure 5.8 Energy curves of bent 34-walled nanotube with respect to the bending curvature. Each mark represents the strain energy for fictitiously uniform bending (square), the strain energy for actually rippled deformation (cross), and the total energy (i.e., sum of the strain energy and the vdW energy) for rippled deformation (circle). Reprinted with permission from Arroyo, M., and Belytschko, T., *Phys. Rev. Lett.*, **91**, 215505-1–4, 2003. Copyright (2003) by the American Physical Society.

advantageous to release much of the membrane strain energy at the expense of slight flexural energy. As a result, rippled MWNTs have a significantly lower strain energy than uniformly bent MWNTs.

Figure 5.8 shows the energy of bent MWNTs as a function of the bending curvature for the 34-walled nanotube. When uniformly bent, the strain energy grows quadratically with respect to the curvature. For such a uniform bending, the vdW energy gives almost no contribution to deformation, and therefore the total energy also follows the quadratic law. However, the actual behavior of the system greatly deviates from this ideal linearly elastic response. As can be observed in Fig. 5.8, the rippling deformation leads to much lower values of strain energy and an increase in vdW energy. The evolution of the total energy E_{tot} with respect to the curvature radius R is very accurately fitted by $E_{\text{tot}} \propto R^{-a}$ with $a = 1.66$; this response differs from that predicted by atomistic simulations of SWNTs or small hollow MWNTs, with an initial nearly quadratic growth ($a = 2$), followed by a postbuckling regime characterized by a linear growth ($a = 1$). These results evidence the failure of the linear elasticity

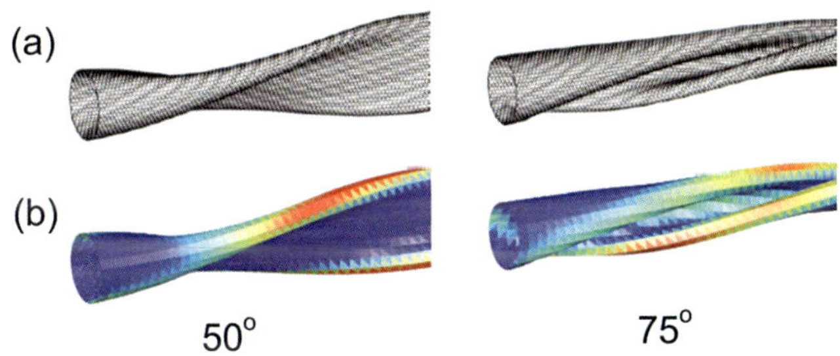

Figure 5.9 Simulated results of twisted SWNTs for two twisting angles. The strain energy distribution is depicted by color; red is high, blue is low. Reprinted with permission from Zou, J., Huang, X., Arroyo, M., Zhang, S., *et al.* (2009), Effective coarse-grained simulations of super-thick multi-walled carbon nanotubes under torsion, *J. Appl. Phys.*, **105**(3), 033516-1–8. Copyright 2009, American Institute of Physics.

and linearized stability analysis to explain the observed well-defined post-buckling behavior ($1 < a < 2$), implying the need of a new theoretical framework based on the non-linear mechanics.[a]

5.5 Twist Buckling

5.5.1 *Asymmetric Response*

Compared with the extensive works on nanotube buckling under bending, relatively few studies have been done on their torsional buckling. In principle, SWNTs under torsion exhibit a sudden morphological change at a critical torque in similar to bending situations, transforming into a straight axis helical shape. The crucial difference from bending cases is that under torsion, the critical buckling torque of SWNTs is dependent in general on the loading direction, i.e., whether right-handed or left-handed manner the tube is twisted [172, 173]. This load direction dependence originates from the tube chirality of chiral SWNTs that breaks the rotational symmetry of the carbon atomic network about the tube axis. This geometric asymmetry is known to markedly influence the

[a] Interestingly, the energetic of MWNT bucking under bending alters by inserting cross-linking (i.e., sp^3 covalent bonding) between adjacent walls, as presented in Ref. [522].

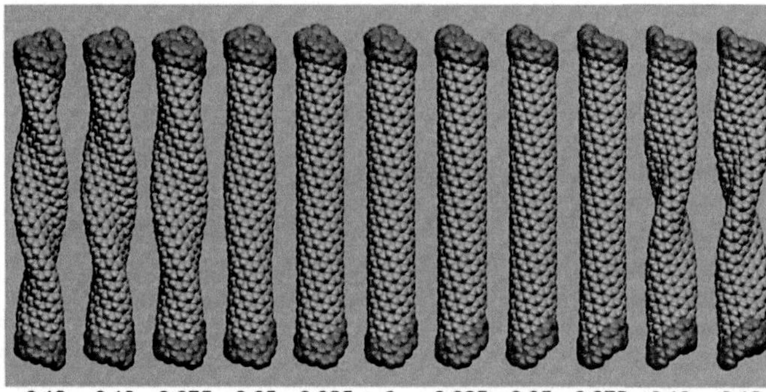

Figure 5.10 Morphological changes for a (8, 3) nanotube under torsion. Reprinted with permission from Chang, T. (2007), Torsional behavior of chiral single-walled carbon nanotubes is loading direction dependent, *Appl. Phys. Lett.*, **90**(20), 201910-1–3. Copyright 2007, American Institute of Physics.

mechanical properties of chiral SWNTs under torsion. For example, the twisting failure strain of chiral SWNTs in one rotational direction may even be 25% lower than that in the opposite direction [173]. Besides, the symmetry breaking causes coupling between axial tension and torsion, giving rise to a axial-strain-induced torsion of chiral SWNTs [174].[a]

The effect of structural details on buckling of a torsional SWNT was explored using MD calculations [172]. Figure 5.10 shows morphology changes of a (8,3) SWNT under torsion. Its torsional deformation is significantly dependent on the loading directions. Under the right-handed rotation, the tube buckles at a critical buckling strain $\gamma_{cr} = 7.6\%$, 1.77 times higher than that ($\gamma_{cl} = 4.3\%$) under the left-handed rotation.[b] Figure 5.11 summarizes a systematic computation [172] of the critical buckling shear strains in both twisting (γ_{cr}) and untwisting (γ_{cl}) directions as function of tube chiral angle. Loading direction-dependent torsional response

[a]This intriguing coupling effect shows that a chiral SWNT is able to convert motion between rotation and translation, thus promising a potential utility of chiral SWNTs as electromechanical device components.

[b]Obviously, it is dangerous if so much a difference between the critical buckling strains in the two loading directions is ignored in designing a nanotube-based torsional component in nanoelectromechanical systems [175–178].

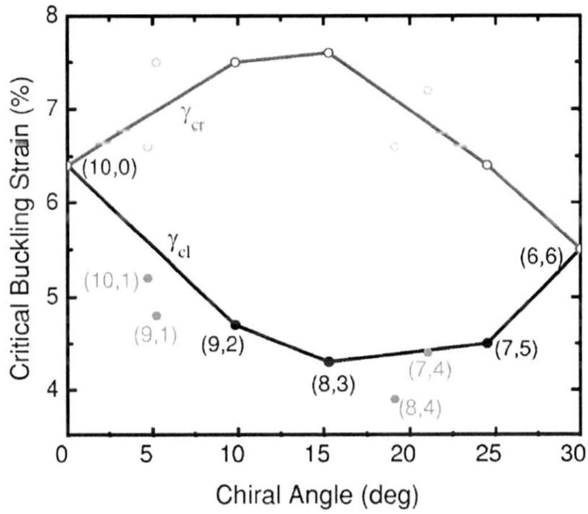

Figure 5.11 Critical buckling shear strains as function of tube chirality for SWNTs with approximately the same diameters. Some additional data for SWNTs with slightly larger or smaller diameters are also presented for reference. Reprinted with permission from Chang, T., (2007), Torsional behavior of chiral single-walled carbon nanotubes is loading direction dependent, *Appl. Phys. Lett.*, **90**(20), 201910-1–3. Copyright 2007, American Institute of Physics.

of chiral tubes is clearly observed. For a given tube diameter, with increasing tube chiral angle, γ_{cr} (γ_{cl}) increases (decreases) from the value for zigzag tubes to the maximum (minimum) value and then decreases (increases) to the value for armchair tubes. Special attention should be paid to that the maximum difference between γ_{cr} and γ_{cl} is up to 85%. This clear difference in the mechanical response suggests particular caution in the use of carbon nanotubes as torsional components (e.g., oscillators and springs) of nanomechanical devices.

5.5.2 *Non-Trivial Responses*

There have been relatively less research investigation on double-walled carbon nanotubes (DWNTs) and MWNTs under torsion because of their relatively complex structures and computational costs. As to twisted DWNTs, MD simulations have revealed [144] a non-trivial buckling mode involving a few thin, local rims on

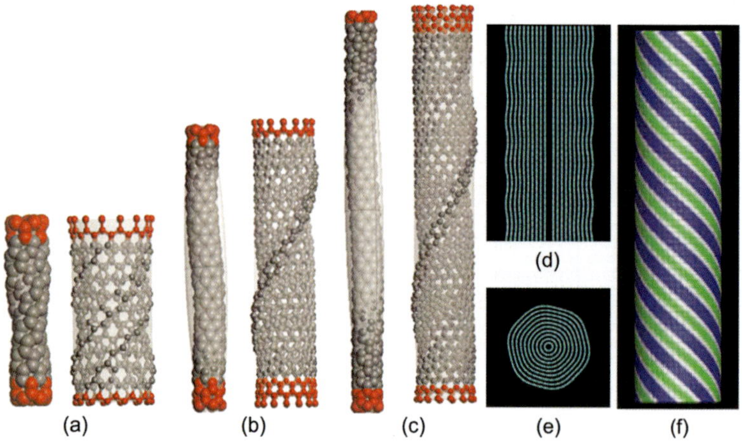

Figure 5.12 (a)–(c) Helical buckling mode in three dimensions of the inner wall and buckling mode with local rims of the outer wall of (5,0)@(14,0) DWNTs with the length: (a) $L = 1.095$ nm, (b) 4.45 nm, and (c) 6.97 nm. (d)–(f) Rippling deformation of a 10-walled nanotube (5,5)@...@(50,50) under torsion with 34 nm in length and 3.4 nm in radius. (d) Longitudinal view. (e) Cross-sectional view. (f) Deformation map: green for ridges and blue for furrows. (a)–(c) Reprinted from Wang, Q. *Carbon*, **46**, 1172–1174 (2008). Copyright (2008), with permission from Elsevier. (d)–(f) Reprinted with permission from Zou, J., Huang, X., Arroyo, M., Zhang, S., *et al.* (2009), Effective coarse-grained simulations of super-thick multi-walled carbon nanotubes under torsion, *J. Appl. Phys.*, **105**(3), 033516-1–8. Copyright 2009, American Institute of Physics.

the outer tube while the inner tube shows a helically aligned buckling mode (see Fig. 5.12(a)–(c)). These distinct buckling modes of the two concentric tubes imply that a conventional continuum approximation in which it is postulated that the buckling modes of all the constituent tubes have the same shape fails for analyzing the torsional responses of DWNTs.

When increasing the constituent walls far more than two, we acquire torsional rippling deformations.[a] It was numerically found that [158] the amplitude in the torsional rippling of MWNTs can

[a]During the torsional rippling, the innermost core tube stores a very high strain energy despite the nearly zero rippling amplitude. Owing to the strong confinement, the strain energy in this layer cannot be released via rippling, which may lead to bond breaking and subsequent brittle crack when the torsional deformation continues to increase [154, 157].

be accurately described by a simple sinusoidal shape function, as confirmed by Figs. 5.12(d) and (f). It is worth noting that the characteristics of the helical rippling morphology in twisted MWNTs are different from that in bent MWNTs, i.e., the so-called Yoshimura or diamond buckling pattern. Structurally, torsion induced rippling is distributed more periodically and uniformly along the tube while bending induced rippling is located only in the compressive side [152, 155]. Energetically, in twisted MWNTs the ridges stay in a higher energy state than the furrows, while in bent MWNTs the strain energy is nearly equally concentrated at the folds (both ridges and furrows).

5.6 Universal Non-Linear Scaling

Interestingly, non-linear elastic response of MWNTs under bending and torsion was found to obey universal scaling laws that consist of two distinct power-law regimes in the energy-deflection relation. Figure 5.13 shows the mechanical response of MWNTs under bending and torsion; the strain energy E versus bending curvature κ or twisting angle Θ is plotted with increasing stepwise the number of walls, n, from $n = 10$ to $n = 40$. All the tested MWNTs exhibit two distinct power-law regimes: a harmonic, linear deformation regime characterized by the exponent $a = 2$ for the relation

$$E \propto \kappa^a \quad \text{or} \quad E \propto \Theta^a$$

indicated by blue lines in Fig. 5.13, and an anharmonic, post-buckling regime with exponent $a \sim 1.4$ for bending and $a \sim 1.6$ for torsion (red lines). The latter non-linear response corresponds to ripples of the graphene walls we earlier discussed.

Figures 5.13(b) and (e) exhibit the data collapse for all the tested nanotubes upon a universal scaling law. This law is characterized by the anharmonic exponent a and by the key quantity ℓ_{cr} that has units of length. In twist situations, it is defined by

$$\ell_{\mathrm{cr}} = \gamma_{\mathrm{cr}} R = \frac{\Theta_{\mathrm{cr}} R^2}{L},$$

where L and R are the length and outer radius of MWNTs considered; Θ_{cr} denotes the critical value of twisting angle at which

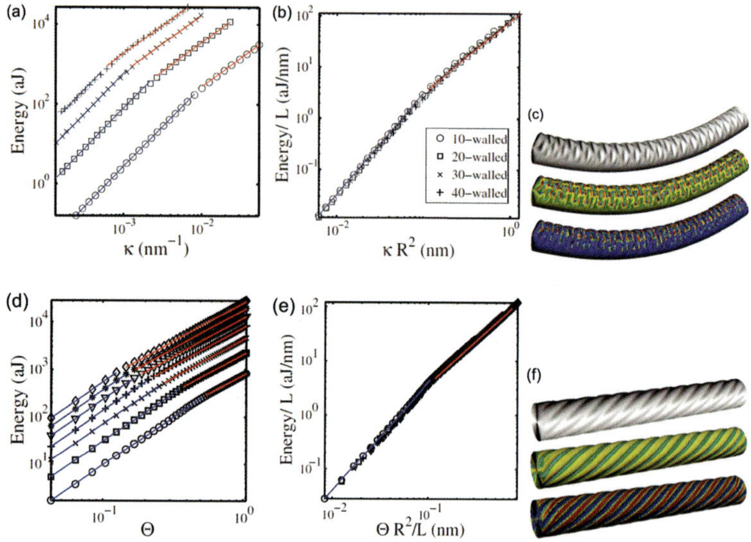

Figure 5.13 Bent MWNTs. (a),(d) Strain energy curve as a function of the bending curvature κ or the twisting angle Θ. (b),(e) Data collapse upon appropriate rescaling. The power law fits with exponents 2 (blue) and $a(< 2)$ (red) are shown for illustration. In all the four plots, the number of walls increases stepwise from 10 (circle) to 40 (cross). (c) 40-walled nanotube in pure bending. (f) 35-walled nanotube in torsion. The latter two panels present deformed shape (top), Gaussian curvature map (middle; green is zero, red is positive, blue is negative), and energy density map (bottom; red is high, blue is low). Reprinted with permission from Arias, I., and Arroyo, M. *Phys. Rev. Lett.*, **100**, 085503, 2008. Copyright (2008) by the American Physical Society.

the buckling arises. In bending cases,

$$\ell_{\text{cr}} = \kappa_{\text{cr}} R^2$$

with κ_{cr} being the critical buckling curvature. Then, the unified law plotted in red and blue in Fig. 5.13(b) and (e) is

$$\frac{E}{L} \propto \begin{cases} (\gamma R)^2 & \text{for } |\gamma R| \le \ell_{\text{cr}}, \\ \ell_{\text{cr}}^{2-a} |\gamma R|^a & \text{for } |\gamma R| > \ell_{\text{cr}}. \end{cases} \tag{5.1}$$

The actual value of ℓ_{cr} was evaluated as $\ell_{\text{cr}} \sim 0.1$ nm for both bending and twisting cases. It should be emphasized that since ℓ_{cr} has dimension of length, the unified law is size dependent; for instance, the thicker the MWNTs, the smaller the critical buckling angle or curvature we will obtain. Such size dependence in materials science is usually associated with irreversible processes, including plasticity or fracture. In MWNTs, instead, it arises as a consequence of a reversible geometric instability.

5.7 Radial Corrugation under Pressure

Added to bending and twisting conditions, radial pressure can yield another interesting class of buckling reflecting the nanotube's high flexibility in the radial direction. In fact, radial stiffness of an isolated carbon nanotube is much less than axial stiffness [179], which results in an elastic deformation of the cross section on applying a hydrostatic pressure. Thus far, many experimental and theoretical studies have been carried out on this issue [94, 180–199]. Most of them focused on SWNTs and their bundles and revealed flattening and polygonalization in the cross section of SWNTs under pressure of the order of a few GPa [182, 200]. Contrary to the intensive studies on SWNTs, radial deformation of MWNTs remains largely behind. Intuitively, the multiple-shell structure of MWNTs is thought to enhance the radial stiffness of MWNTs. However, when the number of concentric walls is much greater than unity, outside walls have large diameters so that external pressure may lead to a mechanical instability in the outer walls. This local instability triggers a novel cross-sectional deformation, called the radial corrugation [201], of MWNTs under hydrostatic pressure, whose nature is intrinsically different from the cases of SWNTs.

Figures 5.14(a) and (b) illustrate an MWNT's cross-sectional view of two typical deformation modes: (a) elliptic ($n = 2$) and (b) corrugation ($n = 5$) modes. In the elliptic mode, all constituent walls are radially deformed. On the contrary, in the corrugation mode, outside walls exhibit significant deformation, while the innermost wall maintains its circular shape. Which mode will be obtained under pressure depends on the number of walls N and the core tube diameter D of the MWNT considered. In principle, larger N and smaller D favor a corrugation mode with larger n.

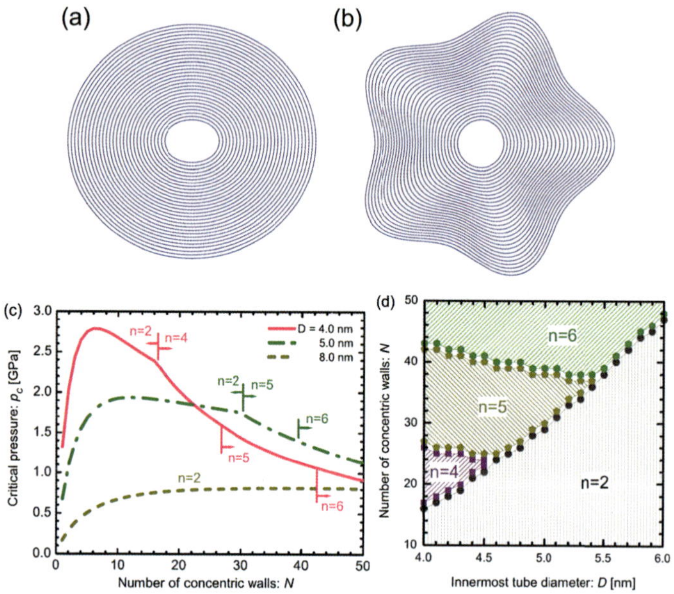

Figure 5.14 (a) Cross-sectional views of (a) elliptic ($n = 2$) and (b) corrugated ($n = 5$) deformation modes observed for N-walled nanotubes with (a) $N = 29$ and (b) $N = 30$. The innermost tube diameter $D = 5.0$ nm is fixed. The mode index n indicates the wave number of the deformation mode along the circumference. (c) Critical pressure curves showing p_c required to produce radial deformation of N-walled nanotubes with fixed D. The mode index n that characterizes the radial deformation mode observed just above p_c is also shown. (d) Phase diagram of radial deformation modes observed above p_c. Various corrugation modes are obtained depending on the values of N and D. Reprinted with permission from Shima, H., and Sato, M. (2008). Multiple radial corrugations in multiwall carbon nanotubes under pressure, *Nanotechnology*, **19**, pp. 495705-1–8.

Figure 5.14(c) shows the critical buckling pressure p_c as a function of N for various values of D. The mode index n of the deformation mode observed just above p_c for fixed N and D is also shown. The increase in p_c in the region of small N is attributed to the enhancement of radial stiffness of the entire MWNT by encapsulation. This stiffening effect disappears with further increase in N, resulting in the decay of p_c. A decay in p_c implies that a relatively low pressure becomes sufficient to produce radial deformation, thus indicating an effective "softening" of the MWNT.

The two contrasting behaviors, stiffening and softening, are different manifestations of the encapsulation effect of MWNTs.

Figure 5.14(d) shows a phase diagram of the radial deformation modes in MWNTs observed above p_c. The bottom region below a chain of solid circles (colored in gray online) corresponds to the elliptic phase, and the top regions surrounded by other symbols indicate the corrugation phases associated with n. It is clearly observed that above the elliptic phase, multiple corrugation modes are formed depending on the values of N and D. It is also observed that smaller D and larger N favor corrugation modes over the elliptic modes. Furthermore, larger N values yield higher corrugation modes with larger n.

Chapter 6

Topological Defects

No man is without faults. The best man is the one with few.
— Quintus Horatius Flaccus, in "Satires"

6.1 Defects in Carbon Nanotubes

Basically, a carbon nanotube consists of a perfect hexagonal network wrapped into a cylinder. In practice, however, it is not as perfect as it seems but contains a certain amount of structural defects. The presence of defects often causes sizeable changes in physical and chemical properties of carbon nanotubes [203, 204]. Therefore, a complete understanding of their generation processes and structural stabilities is highly desirable for carbon nanotube applications as well as in view of academic interest.

Defects in carbon nanotubes are categorized into four groups:

- Topological defect (introduction of non-hexagonal carbon ring)
- Re-hybridization defect (change from sp^2 to sp^3 chemical bonds)
- Incomplete bonding (vacancies, dislocation)
- Impurities with other elements than carbon

Elastic and Plastic Deformation of Carbon Nanotubes
Hiroyuki Shima and Motohiro Sato

ISBN 978-981-4310-75-8 (Hardcover), 978-981-4364-15-7 (eBook)
www.panstanford.com

Among them, this chapter focuses on a class of topological defect. Particular emphasis will be placed on the so-called 5-7-7-5 defects [205] (sometimes called Stone–Wales defects, see Section 6.2). A 5-7-7-5 defect consists of two pentagonal (five-membered) carbon rings adjacent to two heptagonal (seven-membered) rings. This kind of defects are thought to play a primary role in carbon nanotube deformation under large external strains beyond elastic limits. It is also interesting that artificial introduction of topological defects in the hexagonal network allows the synthesis of anomalously shaped carbon nanomaterials with non-tubular geometry.

6.2 Stone–Wales Transformation

A 5-7-7-5 defect, or Stone–Wales (SW) defect [205], is the lowest energy defect in carbon nanotubes. It is formed by a $\pi/2$ rotation of a C–C bond, which transforms four hexagonal carbon rings into two pentagons and two heptagons; see Fig. 6.1. Such $\pi/2$ bond rotation is referred to as a Stone–Wales (SW) transformation. The SW transformation leaves an ensemble of adjacent non-hexagonal (i.e., five- and seven-membered) rings, which is what we call a 5-7-7-5 defect. This defect structure is a metastable state, and thus the system needs to overcome an activation barrier of several eV to realize it.

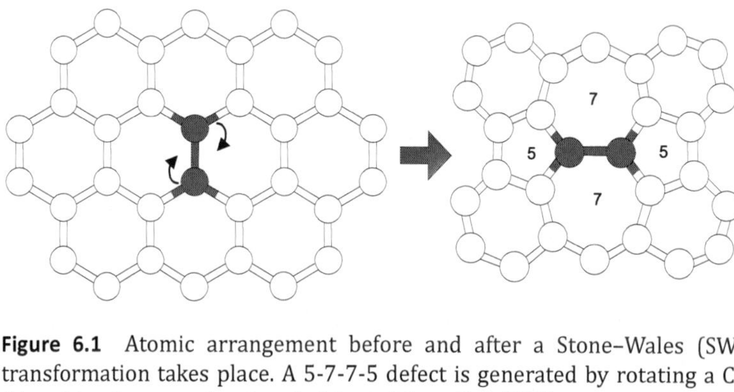

Figure 6.1 Atomic arrangement before and after a Stone–Wales (SW) transformation takes place. A 5-7-7-5 defect is generated by rotating a C–C bond in a hexagonal network. Adapted from Ref. [222].

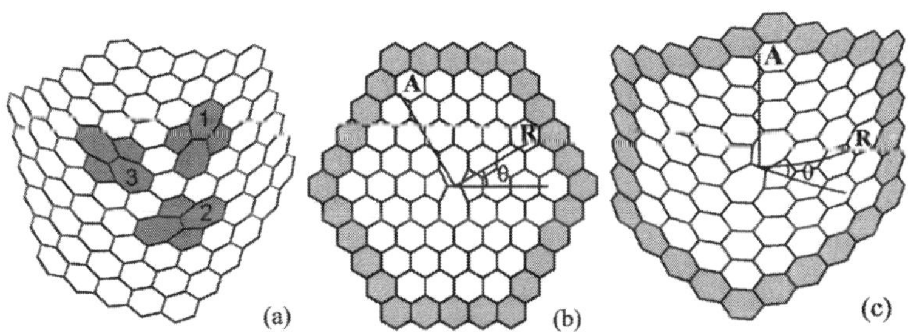

Figure 6.2 (a) Three possible orientations of 5-7-7-5 defects. (b) A 5-7-7-5 defect embedded in a flat graphene sheet. (c) Rolling up the sheet in the ***R*** direction by a curvature radius *r*. The vector ***A*** points to the tube axis, and θ denotes the relative angle of the 5-7-7-5 defect with the direction of ***R***. Reprinted with permission from Zhou, L. G., and Shi, S.-Q. (2003), Formation energy of Stone–Wales defects in carbon nanotubes, *Appl. Phys. Lett.*, **83**(6), 1222–1224. Copyright 2003, American Institute of Physics.

Interestingly, the activation barrier for the SW transformation declines dramatically [206, 207] when an axial strain is applied to the carbon nanotube. The strain-induced lowering of the barrier is caused by the fact that the SW transformation effectively releases the excess strain energy in nanotubes; this release mechanism will be discussed in Section 6.4.1 and 6.4.2 in details. In fact, numerical simulations [46, 206, 208] have shown that the 5-7-7-5 defect structure becomes energetically favorable above the axial strains of 5% in armchair SWNTs and above 12% in zigzag SWNTs. In addition, a catalytic effect of carbon adatoms also reduces the activation barrier, as has been demonstrated by *abinitio* calculations [207, 209].

Under appropriate conditions, a 5-7-7-5 defect can dissociate into two separated 5-7 pairs by an additional SW transformation of neighboring C–C bonds (see Section 8.2) [210]. Further successive transformations cause each 5-7 pair to migrate along a helical path that twines around the tube [206, 210] (see Figs. 8.2 and 8.3).[a] The migration is possible only when a sufficient amount of energy is supplied via thermal excitation and/or axial strain application.[b] The

---

[a]Splitting of 5-7-7-5 defects is energetically favorable for low strains, whereas higher strain leads to defect aggregation [211].

[b]The energy supply can be also achieved by bend deformation [212, 213].

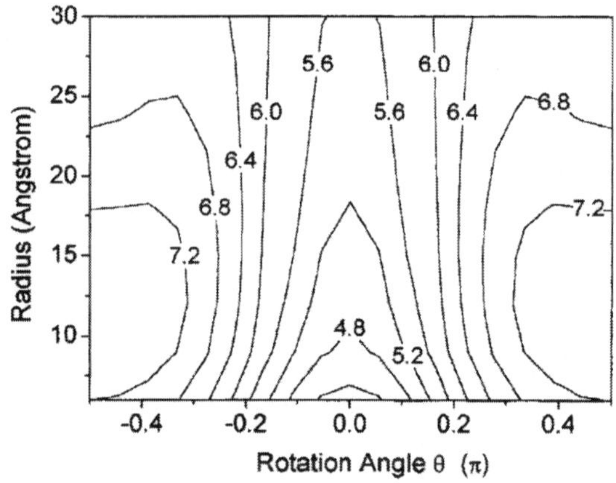

Figure 6.3 Formation energy E_{SW} of a 5-7-7-5 defect in units of eV as a function of the tube radius r and the orientation angle θ defined in Fig. 6.2(c). Reprinted with permission from Zhou, L. G., and Shi, S.-Q. (2003), Formation energy of Stone–Wales defects in carbon nanotubes, *Appl. Phys. Lett.*, **83**(6), 1222–1224. Copyright 2003, American Institute of Physics.

most important consequence of the helical migration of 5-7 pairs is a chirality change of the tube domain that embeds the helical path. Such a chirality change serves as a driving force of a ductile behavior of nanotubes under large axial strain, as has been observed in nanotube elongation that we will discuss in Chapter 8 [210].

6.3 Formation Energy

Let us consider the formation energy of a 5-7-7-5 defect. Given a tube radius r and a chiral angle χ,[a] there are three possible orientations to which a 5-7-7-5 defect is aligned. Figure 6.2(a) shows the three hatched regions at each of which the SW transformation may take place. If we treated a flat graphene sheet, then the three sets are equivalent with respect to the formation energy. However,

[a]The carbon nanotube structure has two degrees of freedom: the tube radius r and the chiral angle χ. The (r, χ) notation can be easily translated from the normal (n, m) notation via the relations: $r = \sqrt{3\left(m^2 + n^2 + mn\right)} \times (a/2\pi)$ and $\tan\chi = \sqrt{3}n/(2m + n)$, where a is the C–C bond length.

this is not the case in carbon nanotubes, since they are not flat sheets but cylinders with small curvature radii around the tube axis. Due to the curvature, an angle between two sp^2 bonds in a nanotube is distorted compared with that in a flat graphene sheet. In addition, the bond lengths perpendicular and parallel to the tube axis are slightly different. Therefore, sp^2 bonds involved in the three hatched regions in Fig. 6.2(a) are non-equivalent,[a] which results in an orientational dependence of the formation energy of a 5-7-7-5 defect.

Defect orientations are described by a variable θ that represents the angle between the direction of the short axis of a 5-7-7-5 defect and the rolling direction of the graphene sheet; i.e., direction $\boldsymbol{R}$ in Fig. 6.2(b). The angle θ ranges from $-\pi/2$ to $\pi/2$, while the chiral angle χ is limited to being $0 \leq \chi \leq \pi/6$. When r and χ are fixed, θ can take a value of $\pi/3 - \chi$, χ, or $\pi/3 + \chi$. Particularly for a zigzag tube with $\chi = 0$, therefore, two out of the three orientations are equivalent. A similar situation arises in armchair tubes with $\chi = \pi/6$.

The formation energy of 5-7-7-5 defects at different r and θ can be calculated by (1) embedding a 5-7-7-5 defect in a flat graphene sheet (Fig. 6.2(b)) that is rotated by θ, (2) rolling the sheet to mimic a part of the tube wall (Fig. 6.2(c)) and relaxing it until mechanically stabilized, and (3) calculating the total energy of the rolled sheet with (E_{def}) and without (E_{perf}) a defect embedded. Finally, we obtain the difference

$$E_{\text{sw}} = E_{\text{def}} - E_{\text{perf}} \tag{6.1}$$

which quantifies the formation energy of a 5-7-7-5 defect. Equation (6.1) implies that a positive formation energy ($E_{\text{sw}} > 0$) makes the 5-7-7-5 defect unfavorable, and vice versa.

The contour plot of E_{sw} on the θ-r plane is shown in Fig. 6.3. When the tube radius r is large enough, E_{sw} converges to that for a flat graphene sheet regardless of θ, as expected. An important observation in Fig. 6.3 is that for a given tube radius, the smaller θ gives the lower E_{sw}. This result means that a C–C bond parallel

---

[a]In precise terminology, the cylindrical geometry of nanotubes breaks the planar symmetry of a graphene sheet and thus the π and σ orbital states can mix and form hybrid orbitals that exhibit partial sp^2 and sp^3 character.

Table 6.1 Predicted formation energies of 5-7-7-5 defects in the graphene sheet and nanotubes based on atomic simulations.

Reference	Formation energy: graphene/armchair/zigzag [eV]
Zhao *et al.* [208]	~5.0 / ~ 4.0 (5,5) / ~ 4.0 (9,0)
Nardelli *et al.* [206]	~2.5 / ~ 2.0 (5,5) / unaddressed
Zhou and Shi [214]	unaddressed / ~ 5.3 (10,10) / ~ 4.6 (20,0)
Li *et al.* [215]	~ 3.6 / ~ 3.5 (5,5) / ~ 3.5 (9,0)
Zhang *et al.* [46]	unaddressed / ~ 4.0 (6,6) / ~ 4.0 (12,0)

Source: Data from Ref. [215].

to the tube axis is easily rotated rather than the one vertical to the axis. In other words, a C–C bond that align perpendicular to the tube axis tends to persist its orientation. The smallest formation energy is found in zigzag tubes with $\theta = 0$, and the largest one in armchair tubes with $\theta = \pm\pi/2$.

To date, several attempts based on different computational approaches have been done for evaluating the formation energy of 5-7-7-5 defects [46, 206, 208, 214]. Table 6.1 summarizes existing results. Despite the data spread, they all suggest that the formation energy is much larger than the thermal energy scale at room temperature and thus nanotubes are very stable in their hexagonal equilibrium configurations.

6.4 Strain-Induced Defect

6.4.1 *Atomistic View*

We have seen in the previous section that the formation energy of a 5-7-7-5 defect is too high to produce it under normal condition. This situation dramatically changes when we apply an axial elongation strain to the nanotube.[a] Under the axial strain, the formation energy can show a significant reduction, and eventually it becomes negative beyond a strain of a few percent [46, 206]; this implies

[a]The tensibility of nanotubes up to at least 5% without breaking was experimentally confirmed [216].

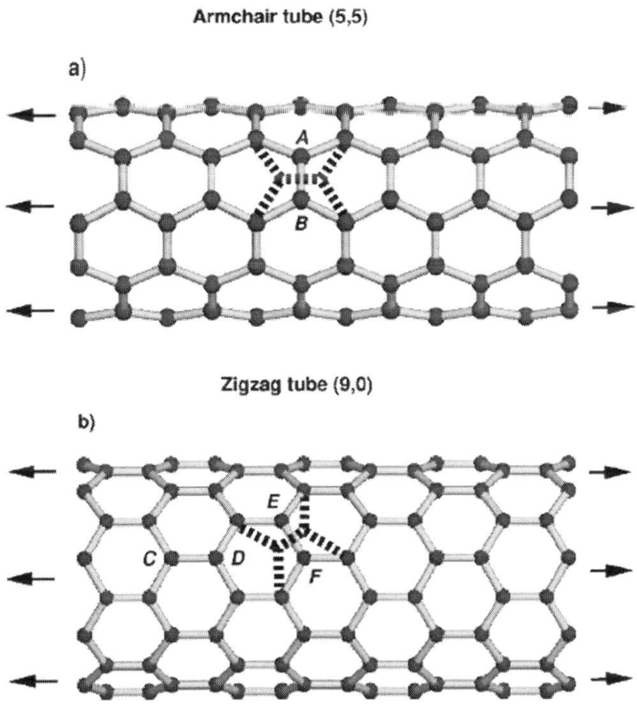

Figure 6.4 Bond rotation in strained carbon nanotubes: (a) side view of a (5,5) armchair tube under axial strain: rotating the bond "AB" by $\pi/2$ forms a 5-7-7-5 defect (dashed lines). (b) Side view of a (9,0) zigzag tube where the bond "EF" is responsible of the formation of the 5-7-7-5 defect (dashed lines). Reprinted with permission from Zhao, Q., *et al.*, Ultimate strength of carbon nanotubes: a theoretical study, *Phys. Rev. B* **65**, 144105 (2002) by the American Physical Society.

the spontaneous generation of 5-7-7-5 defects if nanotubes are subjected to axial strain.

The significant reduction in the formation energy stems from the fact that the SW transformation decreases the applied stress by rotating a C–C bond, which aligns with the direction of the strain. Figure 6.4 shows a schematic of the strain-induced reduction mechanism. In an armchair nanotube, for instance, an axial strain corresponds to a tension applied perpendicular to the bond marked "AB." The tension finds natural release in the $\pi/2$ rotation indicated by dashed lines. As a consequence, the SW transformation effectively elongates the tube in the strain direction, releasing the excess strain energy. In fact, the two heptagons can be stretched more than the hexagons while keeping a C–C bond length that is close to the ideal one.

The above argument implies that different orientations of the carbon bonds with respect to the strain axis lead to different scenarios [206, 211]. This is the case; in a zigzag tube, for instance, the axial tension is parallel to the bond indicated as "CD" in Fig. 6.4(b). Hence, it triggers no rotation of the CD bond. Although the EF bond may rotate by a $\pi/3$ angle under the strain, the resulting 5-7-7-5 defect (indicated by a dashed line) cannot effectively release the excess strain. Owing to these facts, zigzag tubes display a higher strain resistance than armchair tubes with the same diameter [46].

6.4.2 *Energetics*

The formation energies of 5-7-7-5 defects obtained from *abinitio* calculations are shown in Fig. 6.5(a) for both a (5,5) tube (solid line) and the graphene sheet (dashed line). The data for the (5,5) tube decreases almost linearly with strain, and the defect becomes energetically favored at strains above 6%. The same trend is observed for the graphene sheet, although the formation energies are about 1 eV higher than in the (5,5) tube. It is also known that the stability of 5-7-7-5 defects depends on the tube chirality [46, 206, 208]. For example, all armchair nanotubes can have negative formation energy at a large tensile strain [206], while the formation energy does not become negative for $(n, 0)$ zigzag nanotubes with $n > 14$ (i.e., the tube diameter more than 1.1 nm) [211].

Now attention should be paid to the fact that in order to understand the formation mechanism of topological defects in strained carbon nanotubes, one needs to address not only the formation energy, But, equally important, the activation process. While the formation energy is defined by the difference in total energies between a defective and an ideal hexagonal structure (see Eq. (6.1)), the activation energy represents the energy barrier that a system has to overcome to form a defect. If this barrier is sufficiently high, the defect will not be created even if its formation is preferred energetically.

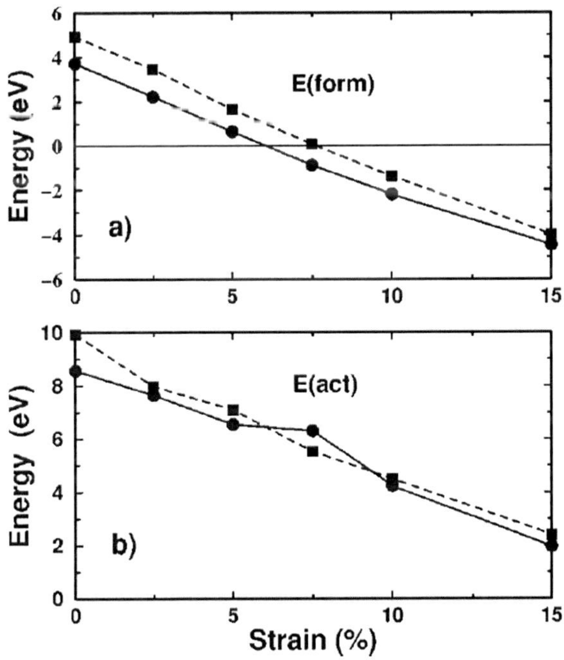

Figure 6.5 (a) Formation energy of the 5-7-7-5 defect in a (5,5) nanotube (circles) and a graphene sheet (squares) as function of uniaxial strain. (b) Activation energy for a SW transformation in the same systems. Reprinted with permission from Zhao, Q., *et al.*, Ultimate strength of carbon nanotubes: a theoretical study, *Phys. Rev. B* **65**, 144105 (2002) by the American Physical Society.

Figure 6.5(b) displays the activation energies for 5-7-7-5 defect formation in both the nanotube and graphetic sheet. These energies are very high: at 0% strain they are 8.6 eV in the (5,5) tube and 9.9 eV in the graphene sheet. Nevertheless, the barriers to defect formation decrease significantly with strain. For example, in a (5,5) tube, it becomes 1.95 eV at 15% strain; this energy scale corresponds to approximately 2×10^4 K in temperature. It is thus concluded that although the topological defects become energetically stable at strains of the order of a few %, the activation barriers for their formation are extremely high, thus hindering the frequent creation of such defects even at large strains. Chirality dependence of the activation energy is also discussed in Ref. [217].

6.5 Direct Micrograph Image

In spite of the intensive theoretical work on 5-7-7-5 defects, their experimental identification has been elusive; somewhat indirect evidence has been provided till recently [218–220]. The main reason is the high activation energy barrier (of a few eV) for the SW transformation [221]. Due to the high activation barrier, the number

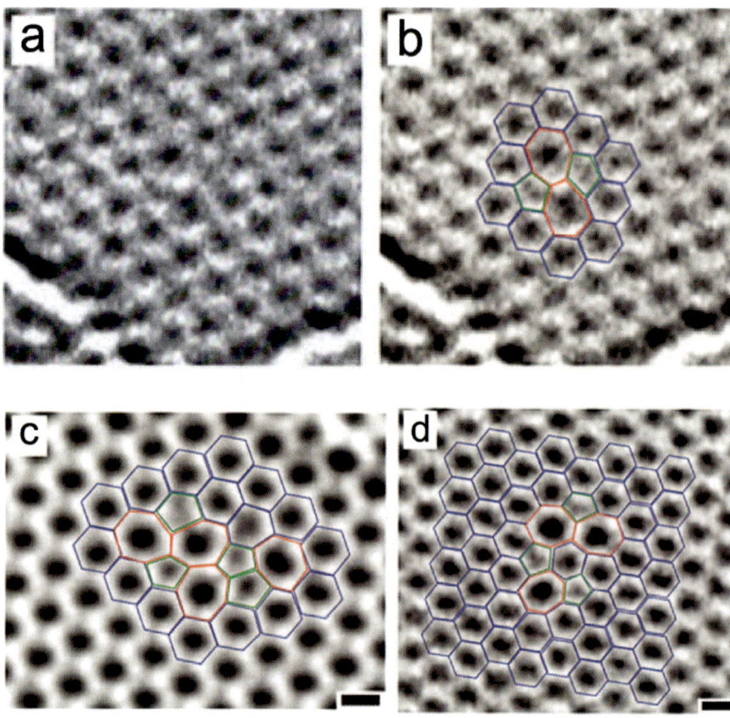

Figure 6.6 (a) Microscope image of a Stone–Wales defect; (b) The same image with atomic configuration superimposed; (c) Defect image and configuration consisting of four pentagons (green) and heptagons (red); (d) Defect image and configuration consisting of three pentagons (green) and three heptagons (red). All scale bars are 2 Å. Reprinted with permission from Meyer, J. C., Kisielowski, C., Erni, R., Rossell, M. D., Crommie, M. F., and Zettl, A. (2008). Direct imaging of lattice atoms and topological defects in graphene membranes, *Nano. Lett.*, **8**, 3582–3586. Copyright (2008) American Chemical Society.

of defects in a nanotube is estimated to be only one defect in several tenths mm at a typical growth temperature (~1400 K). Such slight density of the defects has long prevented a proper identification by ordinary microscope techniques.

The first direct observation of 5-7-7-5 defects was succeeded in 2007 [222] by means of high-resolution transmission electron microscopy (HR-TEM) with atomic sensitivity. Detection of "native" topological defect would be statistically unrealistic; hence, in the HR-TEM measurement, topological defects were artificially introduced in SWNTs by applying high temperature (~2300 K) in vacuum. Such heat treatment led to the fusion of adjacent SWNTs into a unified SWNT with large tube diameter [223]. The latter thick SWNT is obtained by local melting of the constituent thin SWNTs in which a portion of hexagonal lattices is destructed due to thermal excitations. By rapidly cooling the heated specimens, SWNTs having substantial numbers of topological defects were eventually obtained, typically a few defects per 10 nm length.

Shortly after, the same class of defects was observed in a flat single-layer graphene in 2008 [224]. Figure 6.6 shows the obtained HR-TEM image of defects; in particular, the upper two panels provide clear evidence for 5-7-7-5 defects existing in the examined graphene layer. The bottom two panels present images of multiple defect configurations: the left panel shows a defect consisting of four pentagons plus four heptagons, and the right consisting of three pentagons plus three heptagons. In each panel, pentagon and heptagons are colored in green and red, respectively.

6.6 Curvature Generation

Artificial introduction of topological defects in a "controlled" way into carbon nanotubes enables to create special structures that will have other properties than their original tubular forms. For instance, various new structures such as L-, Y-, and T-branches of nanotubes can be obtained, which are in particular of interest as electronic junctions [225]. Other interesting examples are nanocoils [226–229] and bamboo-shaped carbon nanotubes [230–232], whose syntheses are based on the defect incorporation in a regular way.

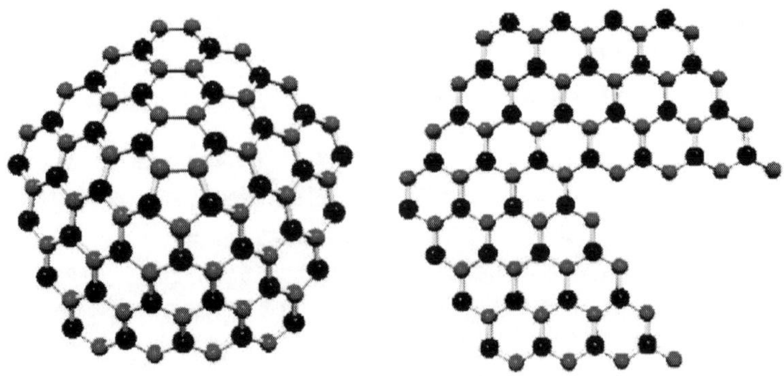

Figure 6.7 Left: Conical surface around a pentagonal defect in a graphene layer. Right: Cut-and-paste procedure to form the pentagonal defect. Removal of the $\pi/3$ wedge from the original flat hexagonal sheet results in a curved surface embedding a pentagonal (non-hexagonal) ring. Reprinted from *Nuclear Phys. B*, **763**(3), Cortijo, A., and Vozmediano, M. A. H., Effects of topological defects and local curvature on the electronic properties of planar graphene, 293–308, Copyright (2007), with permission from Elsevier.

The ability of morphology control by defect insertion originates from the geometric curvature around the defect. To understand it, let us consider the role of five-membered rings in a single graphite layer [233]. Suppose that we prepare a sheet of hexagonal network, remove the $\pi/3$ wedge and then connect the two new edges (see Fig. 6.7). As a consequence, we will obtain a conical surface of a hexagonal network with an apical five-membered ring. In contrast, if we insert an additional $\pi/3$ wedge into either of the two cuts, we obtain a saddle-shaped surface with a seven-membered ring at the center. Similarly, a $2\pi/3$ wedge inserted into the original hexagon sheet will give an octagonal ring embedded in a saddle-shaped surface. Many other structures that contain larger rings (9-, 10-membered rings or more) or smaller rings (four- and three-membered) can be built in principle, although most of them are mechanically unstable [234].

Figure 6.8 provides a manifestation of the defect-induced curvature effect, showing a nanotube cap with spatially varying tube radius. An apex indicated by "P" incorporates a single pentagon

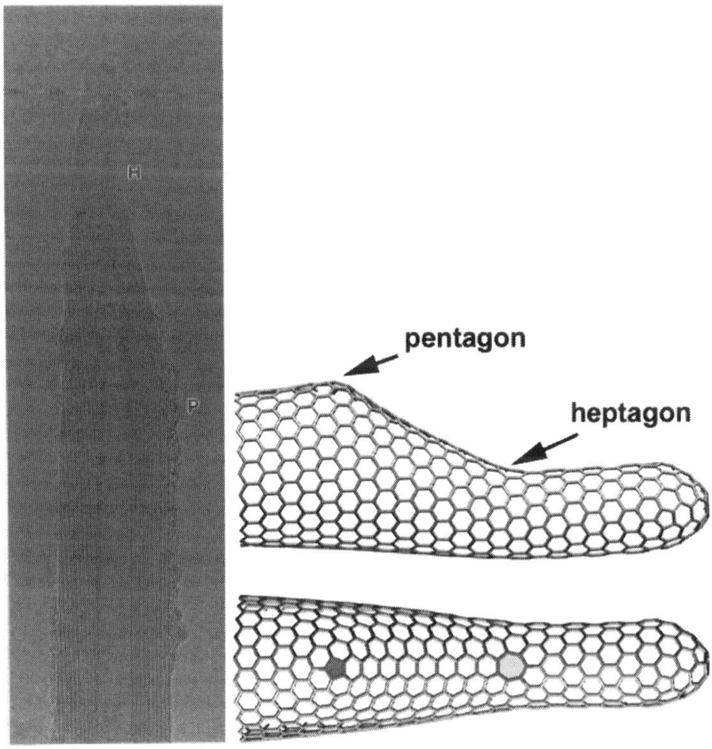

Figure 6.8 Left: An MWNT cap in which the successive presence of a heptagon (H) and a pentagon (P) changes the diameter of the structure. Reprinted with permission from Ebbesen, T. W. (1998). Cones and tubes: geometry in the chemistry of carbon, *Acc. Chem. Res.*, **31**(9), 558–566. Copyright (1998) American Chemical Society. Right: Molecular model of a nanotube containing one pentagon and one heptagon resembling the image shown in the left panel. Reprinted with permission from Terrones, M. (2010). Sharpening the chemical scissors to unzip carbon nanotubes: crystalline graphene nanoribbons, *ACS Nano*, **4**(4), 1775–1781. Copyright (2010) American Chemical Society.

into a hexagon sheet.[a] The opposite situation occurs at the point indicated by "H", where the graphitic shell is folded back to lie in the same direction as in the thicker tube. The hexagon network of this region was obtained by introducing a single heptagon.

[a]Since the introduction of a pentagon results in a strain around the pentagon, the cross section of the tube at P is not perfectly circular but oval with a slight degree.

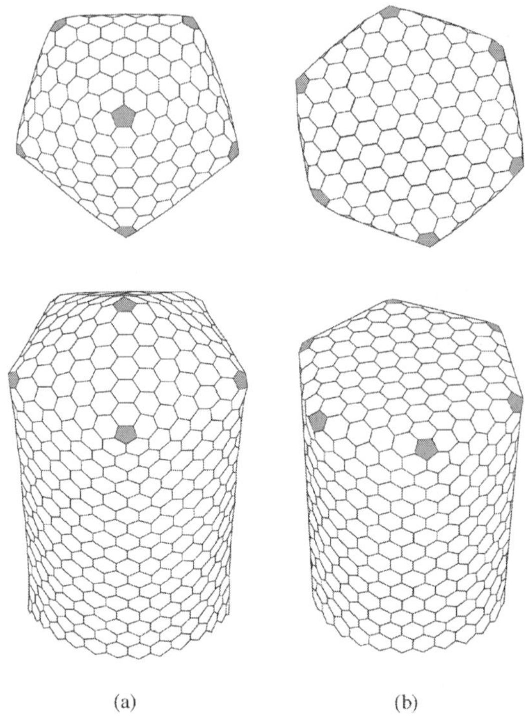

Figure 6.9 Geometries of nanotube caps exhibiting (a) five-fold symmetry and (b) six-fold symmetry. Upper figures are top views, and bottom figures are bird's-eye views. Bodies (sidewalls) of both nanotubes are zigzag type with chiral indices (30,0). Reprinted with permission from Saito, Y., Hata, K., and Murata, T. (2000). Field emission patterns originating from pentagons at the tip of a carbon nanotube, *Jpn. J. Appl. Phys.*, **39**, L271. Copyright 2000 The Japan Society of Applied Physics.

Another manifestation of the defect-induced curvature can be observed at a closed cap at each end of nanotubes. Figure 6.9 shows nanotube caps exhibiting: (a) five-fold symmetry and (b) six-fold symmetry, in which nanotube ends are closed with caps by introducing pentagon rings. Specifically interesting is the fact that as one sees in the figures, precisely "six" pentagons are contained within the cap region in the both cases; other choices of the number of pentagons are prohibited. It should be surprising that this restriction with respect to the number of pentagons holds

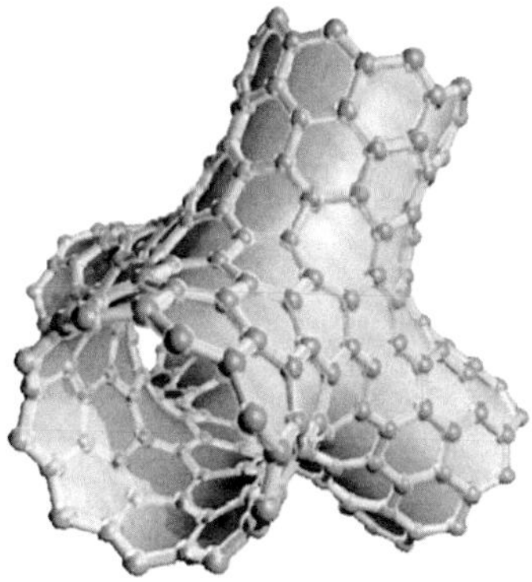

Figure 6.10 The structure of a carbon tetrapod. The four extremities consist of (9, 0) nanotubes with a zigzag edge. Reprinted with permission from Park, N., Yoon, M., Berber, S., Ihm, J., Osawa, E., and Tománek, D., *Phys. Rev. Lett.*, **91**(23), 237204, 2003. Copyright (2003) by the American Physical Society.

for arbitrary tube radius and chirality. In fact, the restriction is a consequence of a mathematical theorem (called Gauss–Bonnet's theorem) that relates the number of non-hexagonal rings to the curvature of the embedding hexagonal surface.

Besides of the closed shapes depicted in Fig. 6.9, a three-dimensional open structure based on a "tetrapod" geometry (Fig. 6.10) is possible. The tetrapod is formed by a network of hexagons and heptagons only and can be regarded as a nanotube junction of four (9, 0) nanotube extremities. This is a possible building block of the recently synthesized nanostructured carbon foam [235].

The last example of defect-induced structures is exceptional in the sense that non-hexagonal rings give no curvature in the embedding hexagonal surface [236]. It is called the topological line defect, defined as a boundary between the armchair and zigzag lines aligned in parallel. Such a topological line defect can be a sequence of alternating pentagons and heptagons (left panel of Fig. 6.11), of pentagon-heptagon pairs (center panel), or of alternating pentagon pairs and octagons (right panel). In these structures, pairwise configurations of n-membered rings with $n < 6$ and $n > 6$ cancel

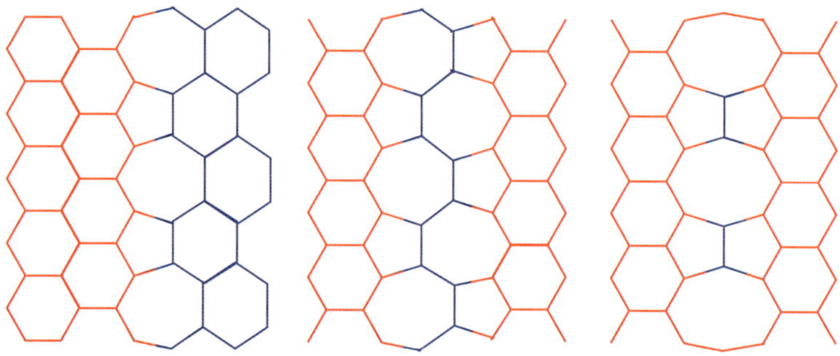

Figure 6.11 Topological line defects defined as a boundary between the aligned armchair (in blue) and zigzag (in red) lines. Left: An alignment of armchair lines adjacent to that of zigzag lines. Center: Two armchair lines inserted into an alignment of zigzag lines. Right: A single armchair line inserted into an alignment of zigzag lines. Adapted from Nakada, K., *et al.*, *J. Math. Chem.*, **45**, 460–470, (2009).

out the positive and negative curvatures that are supposed to be generated around the defects.

6.7 Fusion of Adjacent Nanotubes

An interesting consequence of successive SW transformations is fusion of two adjacent carbon nanotube. Namely, a successive C–C bond rotations allows for two adjacent nanotubes to gradually merge into one nanotube with larger diameter. The propagation of the fused region proceeds in a morphology reminiscent of a Y-shaped junction shown in Fig. 6.12(a). The structure resembles the shape of pants, composed of a (10, 10) nanotube section representing the waist and two separated (5, 5) nanotubes representing the legs. The junction region forms a saddle-shaped surface, at which four heptagons and one octagon are inserted in the hexagonal network.

The fusion proceeds through a periodic sequence of bond rotations. One transformation cycle is depicted in Fig. 6.12(a) from the front view. The detailed bond rearrangements are displayed in Fig. 6.12(b) as structural diagrams. Each step within the transformation cycle is identified by a rotation of a single bond

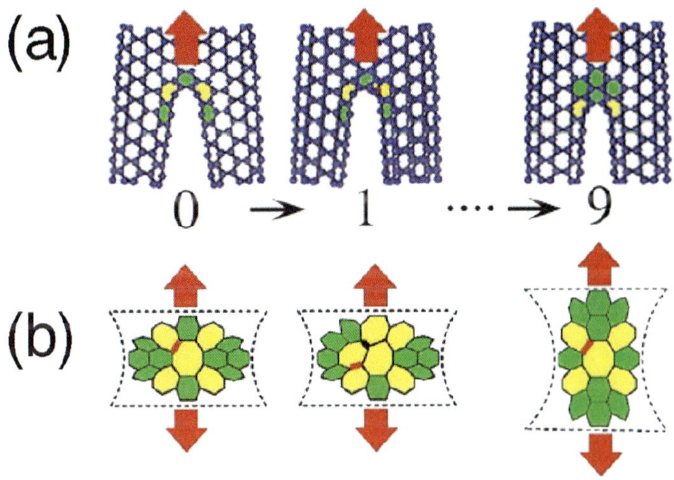

Figure 6.12 Fusion of two adjacent nanotubes that form the Y-junction geometry. One octagon and four heptagons are contained in the junction area. (a) Front view illustrating that two (5, 5) nanotubes merge to a (10, 10) nanotube in a morphology reminiscent of pants. (b) Structural diagrams detailing the bond switching. The fusion process requires nine generalized Stone–Wales (GSW) transformations to propagate the branching region axially by one period of the armchair nanotube. Bonds that are rotated in each step are emphasized by thick lines. Reprinted with permission from Yoon, M., Han, S., Kim, G., Lee, S. B., Berber, S., Osawa, E., Ihm, J., Terrones, M., Banhart, F., Charlier, J.-C., Grobert, N., Terrones, H., Ajayan, P. M., and Tománek, D., L., *Phys. Rev. Lett.*, **92**(7), 075504-1–4, 2004. Copyright (2004) by the American Physical Society.

near the junction; the respective bonds are marked by a thick gray (red) line prior and a thick black line after the rotation. A sequence of nine bond-rotation steps makes the junction region propagate downward along the axis of the (10,10) tube and thus increases the waist section at the expense of the leg section. Following the full transformation cycle, the final structure (labeled by "9" in the figure) is topologically equivalent to the initial structure ("0").

Figure 6.13 shows an experimental evidence of the above-mentioned fusion process [221]. In the measurement, two adjacent tubes are laid in a plane and subjected to electron irradiation at 1100 K in order to secure a sufficient amount of SW transformation. In Fig. 6.13(a), we can distinguish two nanotubes, lying close

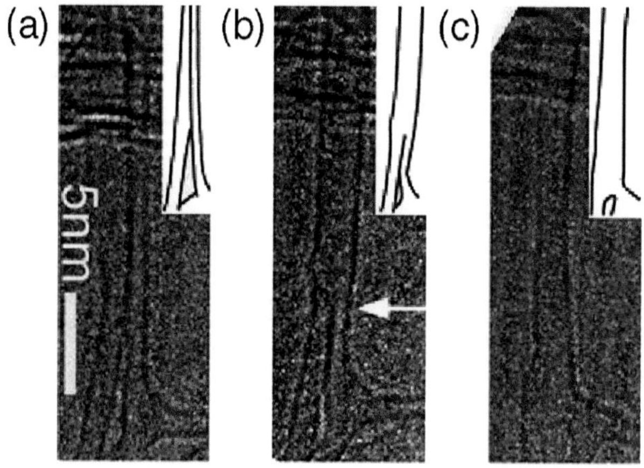

Figure 6.13 Time sequence of high-resolution TEM images evidencing the fusion process for the coalescence of two SWNTs. (a) Tubes prior to coalescence at 1100 K. (b) As electron irradiation continues, the nanotubes get closer, overlap, and start merging into one. The fused region is emphasized by the white arrow. (c) The coalescence process is completed, as the fusion region moves down. Reprinted with permission from Yoon, M., Han, S., Kim, G., Lee, S. B., Berber, S., Osawa, E., Ihm, J., Terrones, M., Banhart, F., Charlier, J.-C., Grobert, N., Terrones, H., Ajayan, P. M., and Tománek, D., L., *Phys. Rev. Lett.*, **92**(7), 075504-1–4, 2004. Copyright (2004) by the American Physical Society.

and parallel to each other. A few minutes later, the tubes were observed to approach even more and started overlapping, as seen in Fig. 6.13(b). Subsequently, the nanotubes formed a local connection and started merging fast. Figure 6.13(c) depicts the final stage of this process, a single wide nanotube. The observed fusion process of two tubes into one tubule with a larger diameter, including the intermediate stages and the speed of interconversion, is consistent with the preceding theoretical arguments.

Chapter 7

Fracture

Hard with hard never made good wall.
— Proverb

7.1 Failure Strength

The failure strength (sometimes called the fracture strength, breaking strength, or ultimate strength) is a stress that should be applied on a given material to break it. An associated quantity is the failure strain, which measures how much strain will be sustained before failure sets in.

For bulk materials, the failure strength and strain can be deduced from a simple uniaxial tensile test. A sample is loaded in a fixed direction until it breaks out and the resulting stress–strain curve tells us the failure strength in addition to other mechanical properties such as Young's modulus and yield strength[a] (Fig. 7.1). As to nanotubes, however, it is fair to say that experiments probing

[a]A yield strength is defined by the stress at which a material begins to deform plastically. Prior to the yield point, the material will deform elastically and will return to its original shape when the applied stress is removed. Once the yield point is passed, some fraction of the deformation will be permanent and non-reversible.

Elastic and Plastic Deformation of Carbon Nanotubes
Hiroyuki Shima and Motohiro Sato

ISBN 978-981-4310-75-8 (Hardcover), 978-981-4364-15-7 (eBook)
www.panstanford.com

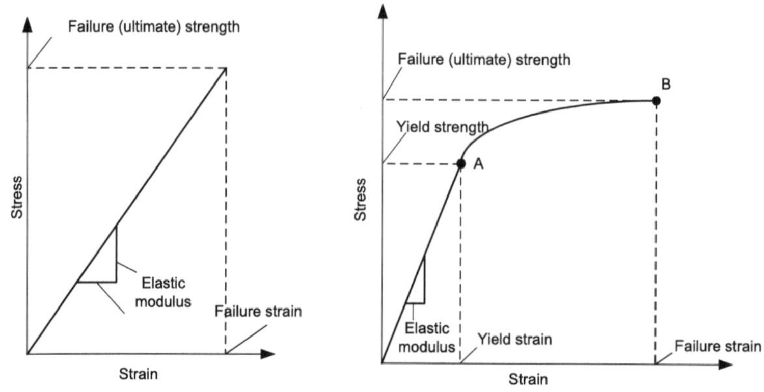

Figure 7.1 Stress–strain curves for a brittle material (left panel), and a ductile material (right). The curves illustrate how mechanical properties of a material can be determined. Ductile materials (such as metals) start to exhibit plastic deformation at the yield point A, before eventually failing at the failure point B. Differently, brittle materials (like ceramics) fail without going through a plastic deformation phase. Adapted from Stach, E. *Nat. Nanotechnol.*, **3**, 586–587 (2008).

their failure strengths are very challenging, due to the difficulties in accurate measurement of the stress that exerts on nanoscale objects as well as in growing high-quality, defect-free nanotubes of sufficient length.

7.2 Tensile Load Measurements

The first experiment aiming at measuring the failure strength and strain of nanotubes was conducted in 2000 [29, 30] for MWNTs produced by arc-discharged methods. Tensile-loading tests were carried out by attaching MWNTs to two opposing atomic force microscope (AFM) tips and pulling them apart until failure was observed.[a] The same test was performed successively for 19 different MWNTs. As a consequence, all the tubes were observed to break in the so-called sword-in-sheath mechanism; failure was

[a] The technical difficulties in carrying out these experiments are illustrated by the fact that in about 50% of the attempted load experiment [29], the nanotube detached itself from one of the two tips before failure was observed.

restricted to the outer wall of the MWNTs (i.e., the wall attached directly to the AFM tips), while the inner walls were unaffected and slid out as a sword pulled from its sheath once the outer tube failed. The failure strain ranged from 3% to 12%, and the strength from 11 to 63 GPa. The large scattering in the obtained values reflects different contents in number of structural defects in the outer shells of the MWNTs.

Similar experiments were conducted on bundles of SWNTs [30]. Like in the case of MWNTs, the load was essentially transmitted only to the tubes lying on the outer side of the bundle; hence, the inner tubes remained largely intact and simply slid against the broken outer tubes. The average values of the failure strength and strain, respectively, were found to be 30 GPa and 3.1%.

7.3 Discrepancy Resolved

The experiments mentioned in the previous section were elegantly conceived by themselves. Nevertheless, the interpretation of the results was complicated due to the inability to observe the internal structure of the tubes during the testing. In addition, the obtained failure strengths (around a few tens GPa) [29, 237–239] may not fully relieve us since they are significantly lower than the theoretical predictions suggested prior to the experiments. For instance, quantum mechanics calculations [240–242, 244, 245] have indicated that defect-free SWNTs possess failure strengths of 75–135 GPa and failure strains of 15–30% depending on chirality. Those calculations assumed defect-free nanotubes with perfect hexagonal structures; therefore, the experimentally observed poor strengths are presumably because of the defective nature of the as-grown nanotube structures.

The discrepancy between theory and experiments was partly resolved by a sophisticated testing methodology [246]. The method employed there was based on image-processing application to the transmission electron microscope (TEM) images of fractured nanotubes. It enables the detection of the number of fractured shells during tensile testing with the atomic scale. Hence, one can quantify precisely what magnitude of stress exerts on which parts of

Table 7.1 Experimentally measured strength of various quasi-one-dimensional nanomaterials.

Material	Multi-walledness or wire diameters (nm)	Measured strength (GPa)	Reference
Carbon nanotube	Multi-walled	30	Yu *et al.* [30]
Carbon nanotube	Multi-walled	97–110	Peng *et al.* [246]
WS_2 nanotube	Multi-walled	3.8–16.3	Kaplan-Ashiri *et al.* [246]
ZnO nanowire	30	7	Wen *et al.* [247]
Si nanowire	100–200	12	Hoffmann *et al.* [248]
Ag nanowire	16.5	7.3	Wu *et al.* [249]
Au nanowire	40	5.6	Wu *et al.* [250]

Source: Data from Ref. [523].

nanotubes and thus can obtain convincing stress–strain curves. The work in [246] also presented data comparison between the stress–strain curve experimentally obtained and the existing theoretical predictions. The agreement between the experimental curve and quantum-mechanical simulations is fairly good, despite a noticeable difference in the failure strain that can be rationalized by the presence of small defects. In fact, it was numerically indicated that [244] even a single vacancy defect is sufficient to reduce failure stresses in a (10,0) tube from 124 to 101 GPa and to reduce failure strains from ~20% to ~13%. Thus, the experimental measurements reported here are entirely consistent with the maximum anticipated properties of carbon nanotubes, providing the simple confirmation that MWNTs can have failure strengths and strains near those that have been predicted by quantum mechanical simulations.

Table 7.1 gives a comparison in mechanical strength between carbon nanotubes and other nanoscale objects known as high-strength materials [523]. These data are either yield or failure strengths measured from tension, compression, or bending experiments. Evidently, the strength of carbon nanotubes is well beyond those of other high-strength nanomaterials.

7.4 One-Atom Vacancy Effect

As repeatedly mentioned, the failure strength of actual carbon nanotubes tends to show lower values than the theoretical predictions.

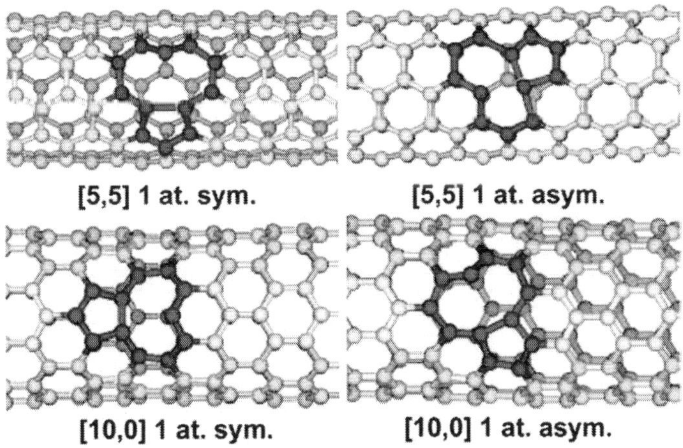

Figure 7.2 The reconstructions of one-atom vacancies in armchair (5,5) and zigzag (10,0) nanotubes. Reprinted from *Chem. Phys. Lett.*, **390**(4–6), Mielke, S. L., Troya, D., Zhang, S., Li, J.-L., Xiao, S., Car, R., Ruoff, R. S., Schatz, G. C., and Belytschko, T., The role of vacancy defects and holes in the fracture of carbon nanotubes, 413–420, Copyright (2004), with permission from Elsevier.

This discrepancy is thought to originate from structural defects pre-existing in the nanotubes. However, it is difficult to clarify these defects experimentally at the atomic level. This is why numerical simulations have been employed to examine the effects of plausible defects on the failure strengths [251].

A primary class of defects that are assumed to pre-exist in nanotubes is a one-atom vacancy. Typical atomic configurations around a one-atom vacancy are displayed in Fig. 7.2. The left two configurations can be named as "symmetric," and the right two are "asymmetric" as intuitively understood from the figures. When a single carbon atom is removed from the hexagonal lattice, a 12-membered ring is left and this can reconstruct to a pentagon and a nonagon. For both armchair and zigzag nanotubes, these reconstructions result in either the symmetric or the asymmetric configuration, as depicted in Fig. 7.2.[a]

[a]At high temperatures, the one-atom vacancies can migrate around the tube [252] and possibly combine with other one-atom vacancies on the same tube to form the more stable double-atom vacancy structures, or combine with vacancies on adjacent tubes to form bridged structures [253].

Table 7.2 Failure strengths and strains of pristine nanotubes, nanotubes with axially aligned 5-7-7-5 defects, and nanotubes involving one-atom vacancy defects.

Chiral index	Defect type	Failure strength [GPa]	Failure strain
(10,0)	Pristine	124	0.20
(10,0)	5-7-7-5	115	0.139
(10,0)	One-atom vacancy	101	0.130
(5,5)	Pristine	135	0.30
(5,5)	5-7-7-5	125	0.22
(5,5)	One-atom vacancy	100	0.153

Source: Data from Refs. [244] and [245] .

Table 7.2 lists the failure strengths and strains for vacancy-defected tubes obtained by quantum mechanical calculations [244]. Comparison values for pristine tubes and tubes with axially aligned 5-7-7-5 defects are also given. The calculations indicate that the one-atom vacancies reduce the failures stresses by 14–26% and markedly reduce the failure strains by about a factor of two for armchair (5,5) tubes. Both symmetric and asymmetric configurations give the same contributions to the failure strength (and strain) reductions; this is because under the axial strain, the two specific configurations (i.e., the symmetric configuration for (5,5) tube and the asymmetric one for (10,10) tube) rearrange to the counterpart configuration through the breaking out of the pentagon's edge that is common to the adjacent nonagon. The marked reductions in strength and strain caused by one-atom vacancy are much larger than those caused by 5-7-7-5 defects, indicating that vacancy defects may be the strength-limiting elements for high-quality carbon nanotubes.

7.5 Large-Hole Effect

Our previous discussion on the one-atom-vacancy-induced strength reductions naturally motivates to consider the effect that large-scale vacancies might have on mechanical properties of the whole nanotube. Two possible candidates are immediately come up with: slit-shaped vacancies and hole-shaped vacancies. The hole defects

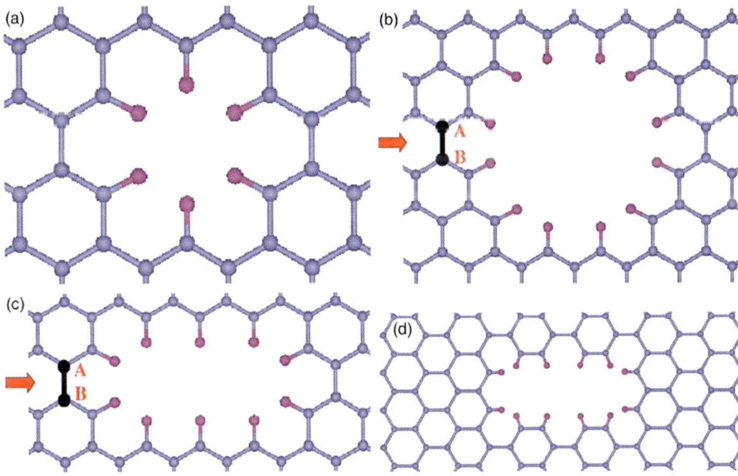

Figure 7.3 Hole and circumferential slit defects in nanotubes (carbon, light blue dot; hydrogen, dark pink dot). (a) A slit or hole of index zero, (b) hole of index one, and (c) slit of index one in a zigzag nanotube; (d) a slit of index one in an armchair nanotube. Local strains are calculated for the bonds in black, shown by red arrows in (b) and (c). Reprinted with permission from Khare, R., Mielke, S. L., Paci, J. T., Zhang, S., Ballarini, R., Schatz, G. C., and Belytschko, T., *Phys. Rev. B*, **75**(7), 075412, 2007. Copyright (2007) by the American Physical Society.

are formed by removing hexagonal units of atoms [244, 251] (see Fig. 7.3), in which the number of removed units denotes the size of the hole.[a] Similarly, slit defects are created by removing rows of carbon atoms the size for a slit is defined so that an n-sized slit is of the same length as the diameter of an n-sized hole.

Figure 7.4 shows quantum-mechanical simulation results of the failure strengths of (50,0) nanotubes with holes and slits. These results show that the failure strengths for holes and slits with the same size differ by less than 5%. Similar agreement between failure strengths of holes and slits was obtained by molecular mechanics [251]. This agreement is in stark contrast to continuum mechanics at the macroscale, where the weakening effect of holes is much less

[a]For example, a zero-sized hole is created by removing one hexagonal unit, a one-sized hole is created by removing a ring of six hexagonal units surrounding the central hexagon, an n-sized hole by removing the ring of hexagonal units surrounding the $(n-1)$-sized hole, and so on.

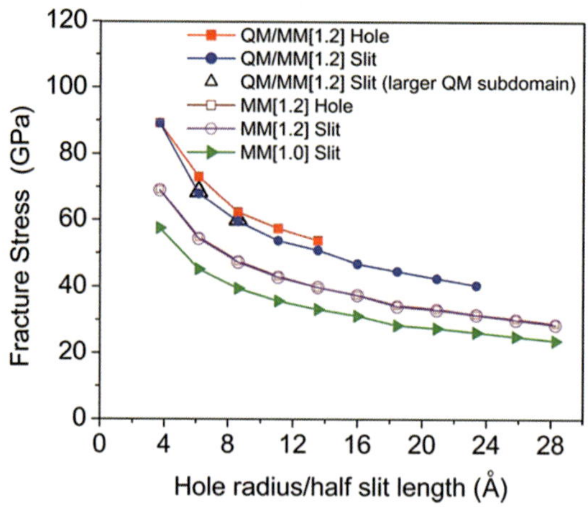

Figure 7.4 Failure strength of a (50,0) nanotube as a function of the defect size for holes and slits. Reprinted with permission from Khare, R., Mielke, S. L., Paci, J. T., Zhang, S., Ballarini, R., Schatz, G. C., and Belytschko, T., *Phys. Rev. B*, **75**(7), 075412, 2007. Copyright (2007) by the American Physical Society.

than the weakening effects of cracks (i.e., slits). Hence, the fracture theory for macroscopic objects cannot be applied to describe the vacancy effects on nanotube mechanics. On the other hand, atomistic simulations are too expensive to solve completely the effect of vacancies, since distributions in sizes, structures, positions of vacancies that are present in one nanotube are diverse. It is, thus, fair to say that the development of breakthrough theoretical approaches would be strongly desirable.

Chapter 8

Super-Elongation

Art is long, life is short.
— Hippocrates

8.1 Candy-Making of a Nanotube

We learned in Chapter 7 that at room temperature, carbon nanotubes typically fail at strains of less than 15%, and in practice only 6% is achieved due to structural defects. Hence, it would be very surprising that at high temperature above 2000 K, an SWNT undergoes superplastic deformation (i.e., super-elongation) to strains of 280% [255]. Furthermore, such exceptional ductile behaviors were observed also in double- and triple-walled carbon nanotubes with an elongation of 190%. Thus, although carbon nanotubes are brittle in normal conditions (as demonstrated in Section 7.3), at sufficiently high temperatures they become extremely ductile.

Figure 8.1 illustrates the super-elongation process of an SWNT. In the actual measurement [255], a piezo manipulator was used to pull an SWNT (24 nm in initial length) to increase the strain. At tensile failure, the SWNT was 91 nm long, showing a tensile elongation of

Elastic and Plastic Deformation of Carbon Nanotubes
Hiroyuki Shima and Motohiro Sato

ISBN 978-981-4310-75-8 (Hardcover), 978-981-4364-15-7 (eBook)
www.panstanford.com

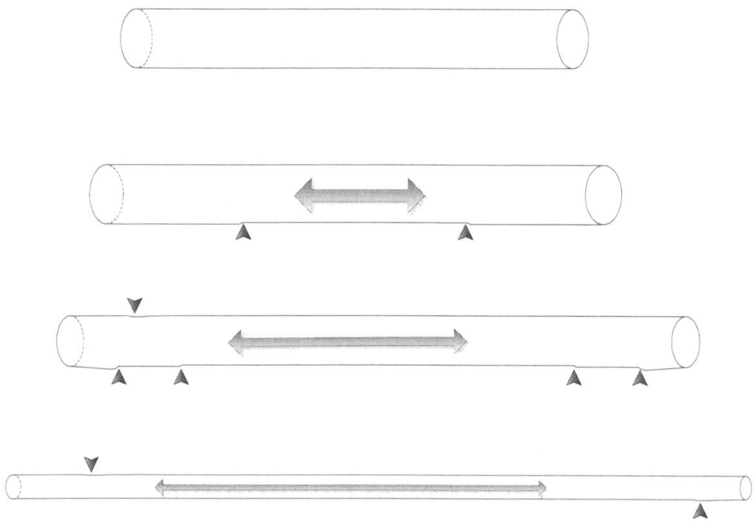

Figure 8.1 Schematic view of SWNT super-elongation under a constant voltage bias. Arrowheads mark kinks that form frequently during tensile straining, propagate along the tube and then pile up or disappear at the ends. Adapted from Ref. [255].

280%; its diameter was reduced 15-fold, from 12 to 0.8 nm. During deformation, the SWNT is heated by more than 2300 K through a contact bias voltage. The same tensile loading without bias (i.e., at low temperature) results in a breakout of the nanotube.

One important clue to understand the 280% tensile strain is the presence of the kinks (a sudden change in the tube diameter) and their motion during elongation. Indeed, several kinks are observed during tensile straining (as drawn in Fig. 8.1), which propagate along the tube and then pile up or disappear at the ends. In addition, the nanotube narrows immediately after the kink passes. These kinks are thought to be associated with topological defects (i.e., 5-7 paired defects) as supported by the theory described in the next section.

8.2 Bond Flip Mechanism

Long before the super-elongation was experimentally observed, theoretical studies had been conducted in order to gain insight into the yield mechanisms of nanotubes under axial strain. The existing theories state that the super-elongation is achieved by the Stone–Wales (SW) transformation, i.e., the nucleation of a 5-7-7-5 defect from the hexagonal graphite network via a $\pi/2$ rotation of a C–C bond (see Section 6.2). At zero stress, the formation of such defects is endothermic. But high-temperature condition along with tensile strain provides external energy to overcome the energy barrier for defect nucleation and motion.

In the following, we outline the mechanism that the strain-induced SW transformations result in the appearance of kinks in nanotubes and allow us to explain the super-elongation phenomena shown in Fig. 8.1. Suppose a (10,10) tube, produced in an almost complete form, involving only one 5-7-7-5 defect (Fig. 8.2(a)). A further SW transformation results in either crack extension (Fig. 8.2(b)) or separation of 5-7 pairs (Fig. 8.2(c)). The former corresponds to the $\pi/2$ rotation of the strictly azimuthal bonds (i.e., the horizontal bond in the figure). It will result in a so-called "7-8-7" defect (by omitting pentagons appearing on both sides of a crack) and further 7-8-8-7, etc., as shown in the bottom of Fig. 8.2(b). This procedure increases the strain of the vertical bonds between these octagons and will eventually lead to their breakage with the formation of larger openings like 7-14-7, etc. The crack represented by this sequence may lead to cleavage.

In a more interesting alternative, the SW transformation of another bond divides a 5-7-7-5 defect into two 5-7 pairs, as a direct inspection of Fig. 8.2(c) shows. The two 5-7 pairs are separated by a single row of hexagons. Next, similar SW rotation results in a double-row-separated 5-7 pairs as shown in the right panel of Fig. 8.2(c). This leads, at high temperatures, to a plastic flow of 5-7 pairs along a helical path in the opposite direction as depicted in Fig. 8.2(d). Indeed, a further simple repetition shows how the 5-7 pairs glide away from each other.

One remarkable feature of such glide is a stepwise change in the chirality of the tube segment bookended by the two separated 5-7 pairs. When the 5-7 pairs sweep a sizeable distance, they leave behind a tube segment changed in the sense of tube chirality. For instance, the formation and splitting of a 5-7-7-5 defect in a (10,10) tube results in a finite segment of (10,9) tube around the

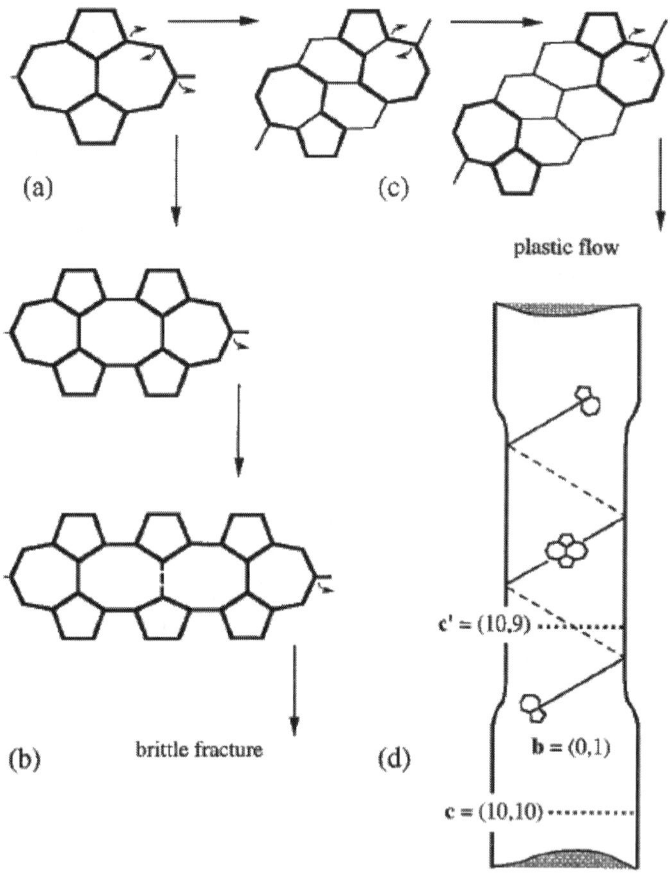

Figure 8.2 (a) In an armchair nanotube, the first SW rotation of an equatorially oriented bond into a vertical position creates a nucleus of relaxation. SW rotations marked by arrows show further evolution as (b) a brittle crack or (c) as a couple of dislocations gliding away from each other. (d) The change of the tube's chirality and a stepwise change of diameter causes the corresponding variations of electrical properties. Formation of a next SW defect continues the necking process, unless the dislocations pileup at insufficient temperature. Reprinted with permission from Yakobson, B. I. (1998). Mechanical relaxation and "intramolecular plasticity" in carbon nanotubes, *Appl. Phys. Lett.*, **72**(8), 918–920. Copyright 1998, American Institute of Physics.

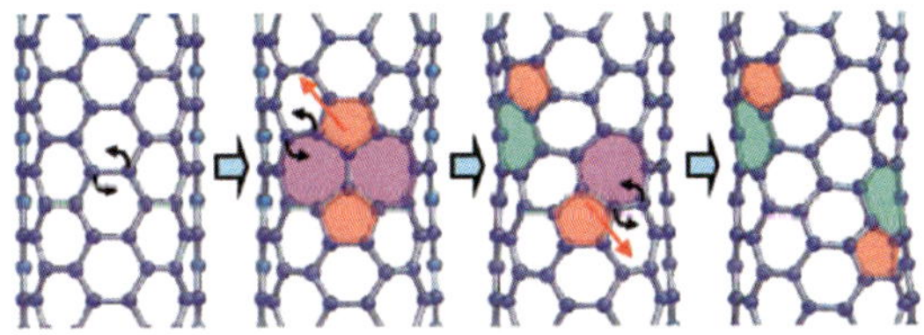

Figure 8.3 (a) Nucleation of a 5-7-7-5 defect in (5, 5) armchair nanotube and its dissociation. (b) SW transformation that causes the 5-7-7-5 defect. (c) Second SW transformation of the neighboring C–C bond. Two 5-7 pairs are separated by a single row of hexagons. (d) Third transformation separates them by a double row. Reprinted with permission from Nakayama, Y. (2007). Plasticity of carbon nanotubes: aiming at their use in nanosized devices, *Jpn. J. Appl. Phys.*, **46**, 5005. Copyright 2007 The Japan Society of Applied Physics.

initial position of the defect generated. Correspondingly, a stepwise reduction in the diameter is observed at the positions of each separated 5-7 pairs. Furthermore, if an additional 5-7-7-5 defect newly appears in the (10,9) segment, then the new two pairs glide away in opposite directions to produce the (10,8) tube segment embedded in the (10,9) segment. By repetition of this process, the (10,10) tube undergoes the following transformation: (10,10) $\to$ (10,9) $\to$ (10,8) $\to \cdots \to$ (10,0) $\to$ [(9,1) or (10,–1)] $\to$ (9,0), etc. This indicates that nanotubes become thin under tensile strain and a high-temperature condition.

The above scenario has shown that armchair tubes display a ductile behavior through the helical migration of 5-7 pairs. These theoretical predictions were also confirmed numerically [213] (see Fig. 8.3); given the axial strain together with high-temperature treatment, successive SW transformations take place, as a result of which the nanotube elongates along the tube axis and shrinks in the perpendicular direction.

8.3 From Brittle to Ductile

We have seen that SW transformations lead to the occurrence of 5-7-7-5 defects that can cause ductile elongation of nanotubes with

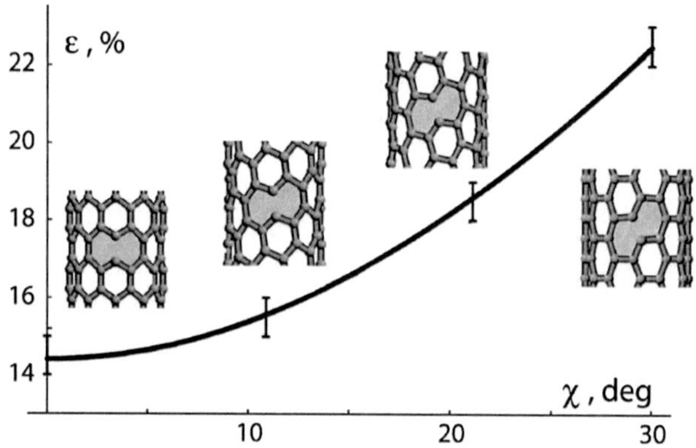

Figure 8.4 Dependence of non-thermal failure strain on chiral angle. At lower points the structures are still stable, whereas the upper strains lead to fracture. Inset: configurations showing the breaking-bond orientations. Dumitrica, T., Ming Hua, M., and Yakobson, B. I. (2006). Symmetry-, time-, and temperature-dependent strength of carbon nanotubes, *PNAS*, **103**, pp. 6105–6109. Copyright (2006) National Academy of Sciences, U.S.A.

the aid of thermal excitation and the axial strain application. On the other hand, nanotubes often exhibit a brittle behavior, which needs no thermal activation but shows a sequence of bond-breaking steps [242, 243]. Figure 8.4 illustrates such non-thermal failure. The failure strain increases modestly with the chiral angle of the nanotubes, as intuitively understood from the correlation between the chirality and bond-breaking orientation depicted in the insets of Fig. 8.4. In the zigzag nanotube with $\chi = 0$, for instance, C–C bonds parallel to the tube axis assume dominantly the axial load, thus breaking at relatively low strain. Meanwhile, in the armchair nanotube with $\chi = \pi/6$, the load is dispersed to the slanted C–C bonds (that make the angle of $\pi/6$ with the axial direction) whose number is twice as much as that of the axially parallel bonds in the zigzag nanotubes. In the armchair case, in addition, the flexibility in the bond angle functions as a cushion effect against the axial load. These facts are the reason why the failure strain in the armchair tube is larger than that of the zigzag tube, as demonstrated in Fig. 8.4.

We should note that Figs. 8.3 and 8.4 show distinct mechanism of nanotube deformation under axial load; the former shows ductile elongation via the SW transformation, while the latter shows

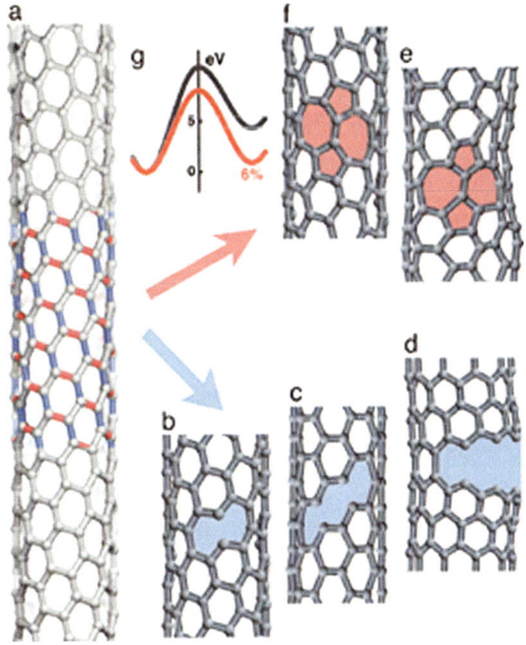

Figure 8.5 Two failure-nucleation processes in nanotubes. (a) A perfect tube under high strain displays nearly longitudinal bonds (blue) more susceptible to direct breaking at low temperature, and those in circumferential direction (red) prone to thermally activated SW rotations. (b–d) Intermediate structures emerging in low-temperature simulations with different numbers of broken bonds n; (b) Chiral tube at $n = 1$; (c) Armchair type at $n = 2$; (d) Zigzag sample at $n = 6$. (e) A 5-7-7-5 defect emerges through SW transformation at high temperature, where the tube is suffered from fluctuating lattice distortion. (f) The defect relaxed at zero temperature. (g) SW flip becomes thermodynamically favored at high tension but its formation still requires high activation, as computed energy profiles at strain $\varepsilon = 0$ and 6% illustrate. Dumitrica, T., Ming Hua, M., and Yakobson, B. I. (2006). Symmetry-, time-, and temperature-dependent strength of carbon nanotubes, *PNAS*, **103**, pp. 6105–6109. Copyright (2006) National Academy of Sciences, U.S.A.

brittle behavior based on the bond breaking. In this context, we may encounter the question as to which is the primary atomic rearrangement and at what strain a nanotube begins to yield in actual experiments.

Figure 8.5 illustrates two main possibilities of local structural changes in nanotubes. Under high tension, the load is transferred differently to the bonds according to their orientations relative to the axis (color-coded in Fig. 8.5(a)). Further, which is the dominant type of atomic rearrangement (i.e., bond-breaking or SW-rotation flip) depends in principle on temperature. At low temperature, the highly elongated bonds (marked blue) breaks and the crack-like configurations emerge (Fig. 8.5(b)). In contrast, at high temperature, thermally excited lattice vibrations promote another bond movement (marked red). The bonds nearly transverse to the axial tension undergo a $\pi/2$ rotation flip into almost longitudinal position (Fig. 8.5(e) and (f)), thus lowering the energy. The formation of a 5-7-7-5 defect becomes favorable under tension but occurs rarely because of its high activation barrier (Fig. 8.5(g)).

8.4 Interdependence in the Failure Mechanism

It is important to realize that the two primary yield mechanisms, brittle crack [242] or ductile deformation [206, 210, 211], are in principle concurrent. The likelihood of their occurrence varies depending considerably on a number of parameters, such as tube diameter, chiral angle χ, applied strain ϵ, test duration t (or similarly, the strain rate $d\epsilon/dt \sim \epsilon/t$), and temperature T. Basically, the brittle mechanism dominates at lower temperatures, while the ductile behavior becomes possible at high temperatures.

Figure 8.6 shows a strength map that accounts for the interdependencies of ε, χ, t, and T. This map indicates that ductile-type bond flip and the brittle bond-breaking mechanisms; that is, either can play the dominant role in failure in a particular test. The yield mechanism is determined by what borderline is crossed first as the tension is applied from $\epsilon = 0$ upward. For instance, at $\chi = 0$ and with increasing ε, we pass across the boundary curve at $\varepsilon \sim 14.5\%$ above which brittle failure based on bond breaking mechanism is likely to occur.[a] Similarly, ductile elongation based on bond rotations

[a]Note that the boundary curve is identical to the curve shown in Fig. 8.4.

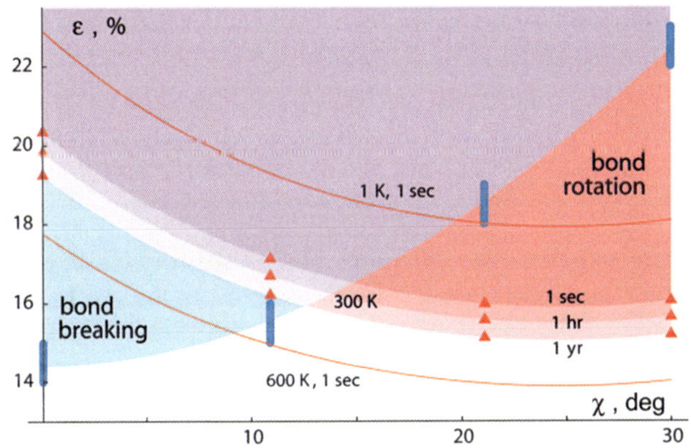

Figure 8.6 Failure-strain map for nanotubes of different symmetry (χ) at various load durations (t) and thermal conditions (T). The yield mechanism is determined by what borderline is crossed first as the tension is applied from $\epsilon = 0$ upward. Brittle-bond breaking for specific tube samples is marked by dark blue segments and extrapolated by blue-domain border; it has no perceptible dependence on T or t. Ductile yield through bond rotation depends on T and t (t = 1 second, 1 hour, 1 year); it was computed for specific tubes (red triangles) and extrapolated to arbitrary tube symmetry (red-level rims). Additional thin red lines show the yield strain at very low (T = 1 K) and higher (T = 600 K) for t = 1 s. Dumitrica, T., Ming Hua, M., and Yakobson, B. I. (2006). Symmetry-, time-, and temperature-dependent strength of carbon nanotubes, *PNAS*, **103**, pp. 6105–6109. Copyright (2006) National Academy of Sciences, U.S.A.

may take place at larger χ when passing across the boundary curve as ε increases.

An important feature of the latter boundary curve is the strong dependence on temperature and time duration. We can see from Fig. 8.6 that the boundary curve translates vertically with varying the time duration as shown by the family of curves (t = 1 second, 1 hour, 1 year) for a fixed temperature of T = 300 K. In addition, the change in T also results in a shift of the curve as illustrated by the family (T =1 K, 300 K, 600 K). The effects of temperature and time duration are significant for the ductile mechanism, but are negligible for the brittle one since it is almost barrierless failure. We also note that both mechanisms of failure nucleation display strong dependence on the nanotube symmetry (χ), with the zigzag type being more at risk to brittle failure [242], whereas the ductile yield is more easily induced in the nearly armchair type.

8.5 C_2 Removal Mechanism

The ductile mechanism accounts for the super-elongation and thinning of nanotube under great external tension. A hallmark of the proposed ductile mechanism is a visible kink composed of the hexagonal lattice warping near the 5-7 paired defect [219]. Following the theory described in Section 8.2, a glide of 5-7 pairs through SW transformations causes stepwise reduction of the diameter and overall elongation of the tube. That is, the kink should propagate along the helical path [210], but cannot move in a purely axial direction since the longitudinal driving force is zero [210]. On the contrary, recent high-resolution microscopy shows a direct longitudinal kink propagation [254, 255] that cannot be attributed to the glide. Furthermore, significant (multifold) loss of the mass was observed [254, 255], which is beyond the glide-based interpretation since SW bond flips are mass conserving.

One possible scenario to explain the longitudinal motion of kinks is based on the removal of two carbon atoms from a particular location. Figure 8.7 illustrates how the removal of C_2 dimers makes the 5-7 pair "climb" up in the "5" direction. Suppose that in

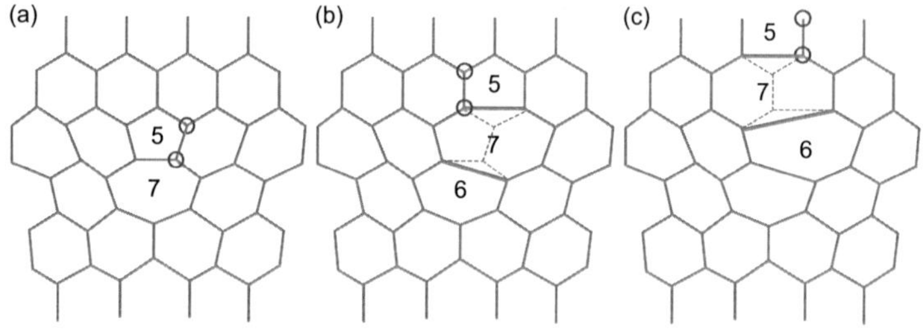

Figure 8.7 (a) A 5-7 pair embedded in the hexagonal lattice. Removal of C_2 atoms marked by the thick circles causes the upward shift of the 5-7 pair as shown in the plot of (b). In a similar manner, the 5-7 pair goes upward again by removing the C_2 atoms marked in (b), as a result of which we obtain the structure depicted in (c).

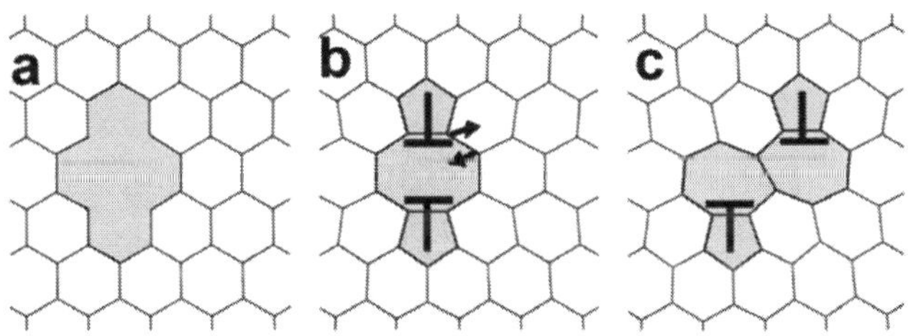

Figure 8.8 (a) Divacancy created by removal of a C_2 dimer. (b) Reconstruction from the divacancy to a 5-8-5 defect. The SW rotation of the C–C bond at the shoulder of the octagon, indicated by arrows, results in the dissociation into two 5-7 defects. (c) The resulting two 5-7 defects. Reprinted with permission from Ding, F., Jiao K., Wu, M., and Yakobson, B. I., *Phys. Rev. Lett.*, **98**(7), 075503, 2007. Copyright (2007) by the American Physical Society.

Fig. 8.7(a), the two atoms marked by thick circles are removed by thermal agitation and/or axial load application. Then, the atomic reconstruction results in new C–C bonds indicated by thick line segments in Fig. 8.7 (b), causing an upward shift of the 5-7 pair. In a similar way, a subsequent removal of C_2 dimer marked in Fig. 8.7(b) leads to the next upward shift as depicted in Fig. 8.7(c). This recurrence procedure describes a new mechanism of longitudinal climb of 5-7 pairs. It is noteworthy that the energy of C_2 removal from the side of the pentagon in the 5-7 core is lower than from anywhere else in the lattice, because it does not create additional dangling bonds or topological defects.

Another interesting property of the removal of C_2 dimers is their role as a source of two 5-7 pairs, complementary to the dissociation of a 5-7-7-5 defect. Figure 8.8 illustrates how the C_2 removal generates two 5-7 pairs. The removal induced divacancy will produce a 5-8-5 defect after relaxation. Then, the SW transformation of the C–C bond at the shoulder of the octagon, indicated by arrows in Fig. 8.8(b), results in the two 5-7 pairs as given in Fig. 8.8(c). Hence, the C_2 removal gives rise to both the generation and translation of the 5-7 pairs, same as the bond flip procedure does as discussed in Section 8.2. In short, there exist two constituent mechanisms that contribute to super-elongation of carbon nanotubes: SW bond flips and C_2 removal.

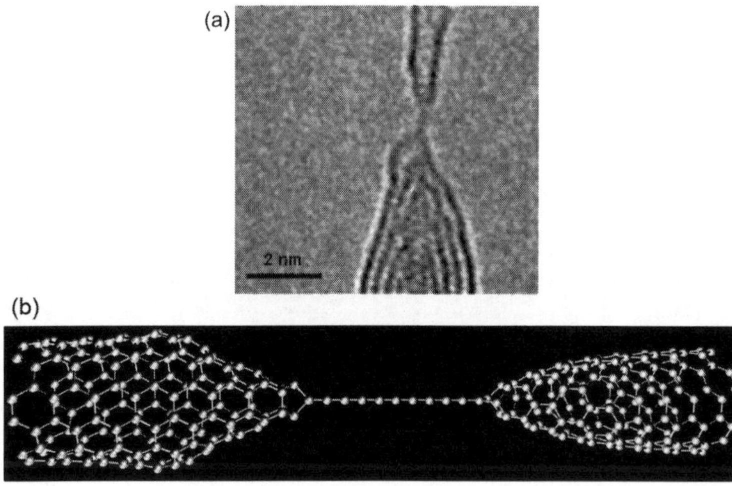

Figure 8.9 (a) TEM image of a mono-atomic C-chain. (b) Typical geometry used in the tight-binding calculations. A chain of up to 9 atoms is placed between two cone-terminated nanotubes. Minimization of these structures leads to a cumulene-type chain. Reprinted with permission from Troiani, H. E., Miki-Yoshida, M., Camacho-Bragado, G. A., Marques, M. A. L., Rubio, A., Ascencio, J. A., and Jose-Yacaman, M., *Nano Lett.*, **3**, 751–755, 2003. Copyright (2003) American Chemical Society.

8.6 Mono-Atomic C-Chain

We conclude this chapter by an overview of an ultimate thinning of nanotube under axial strain, i.e., the formation of a single chain of carbon atoms. Figure 8.9(a) is the TEM image of a monoatomic chain spanning two nanotube fragments before fracture [256]. This was the first direct image by microscope, though similar mono-atomic chains had been experimentally identified in laser-vaporized graphite samples [257, 258] and in the field emission of electrons from biased carbon nanotubes [259].[a]

The formation process and the structural properties of the one-dimensional C-chains can be addressed by extensive tight-binding calculations [261]. Suppose that two finite-size tubes are terminated with cone-like defective structures, and they are bridged by a linear

[a] In the theoretical view, the mono-atomic C chain was predicted for the high-strain-rate fracture of nanotubes [260].

Table 8.1 Tight-binding calculation results of the force constant and length of monatomic C-chain as a function of the number of atoms.

Number of atoms	Force constant [N/m]	Length [nm]
1	81	4.189
2	72	4.325
3	74	4.452
4	66	4.586
5	65	4.713
6	60	4.847
7	56	4.974
8	52	5.107
9	51	5.235

Note: In the length of the structure, it is included the two finite size tubes adjacent to the chain (around 4 nm).
Source: Data from Ref. [256].

chain of carbon atoms (see Fig. 8.9(b)). The high-energetic electron beam (~200 keV) is assumed to drive a random removal of a few carbon atoms in the constriction. In this way, vacancies are created that promote the transfer of carbon atoms from the cone to the chain. Increasing the distance between the tube ends does not break the chain but elongates the linear chain by incorporating atoms from the contact region.

Table 8.1 presents the force constant and length of the mono-atomic chains as a function of the number of C atoms [256]. The equilibrium bond length between carbon atoms lie between 0.127 and 0.134 nm. The bonding is probably of cumulene type (:C=C· · · C=C:), with nearly equivalent bond lengths [262] with only a small dimerization (less than 2%). This is consistent with first-principle calculations for free-standing carbon chains [263]. Chains with less than 10 atoms are predicted to be of cumulene type; however, if the number of atoms increases, the linear configuration becomes unstable. This may be the reason why chains larger than 1 nm were not observed.

Chapter 9

Carbon Nanocoil

Everyone who goes up to a height uses spiral staircase.
— Francis Bacon

9.1 Merits of Coiled Structures

Carbon nanocoils (sometimes called coiled or helical carbon nanotubes) are exotic nanomaterials having coiled geometry, resembling "telephone cords" attached to the phone receiver. They exhibit excellent mechanical properties, which owes to the combination of flexibility due to coil morphology and the toughness resulting from the sp^2 bonding nature. Furthermore, their unique spiral structures imply versatile applications [264] such as resonating elements, nanosprings, and reinforced fibers in high-strain composites. Novel tactile sensors also can be built using carbon nanocoils, which provide very high sensitivity and resolution as high as femtograms [269].

The existence of carbon nanocoils was predicted in 1993 [270], just two years after the boost of carbon nanotube study. Possible structural models for carbon nanocoils were proposed by Ref. [270], together with discussion on the relationships between geometric

Elastic and Plastic Deformation of Carbon Nanotubes
Hiroyuki Shima and Motohiro Sato

ISBN 978-981-4310-75-8 (Hardcover), 978-981-4364-15-7 (eBook)
www.panstanford.com

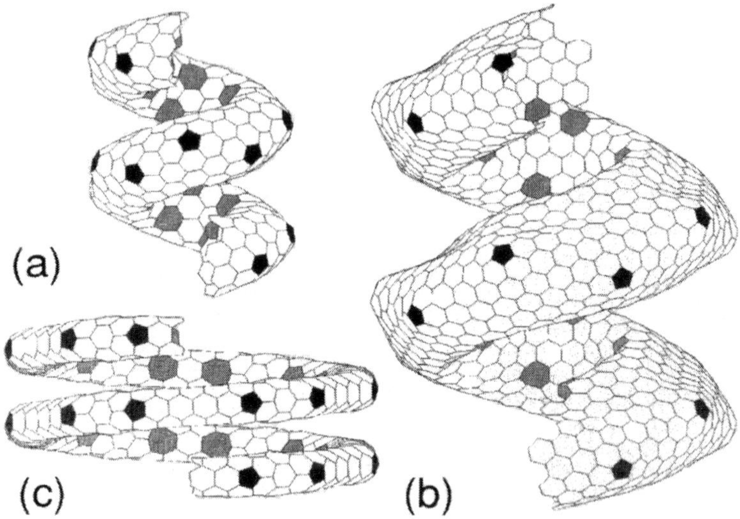

Figure 9.1 Microscopic models of carbon nanocoils predicted by Ref. [270]. Fivefold and sevenfold rings shaded appear in the outer and inner ridge lines, respectively, amid a background of the sixfold rings. Reprinted with permission from Ihara, S., and Itoh, S., *Phys. Rev. B*, **48**, 5643–5648, 1993. Copyright (1993) by the American Physical Society.

parameters (diameter, pitch length, etc.) and physical properties. The coiling of the proposed structure stems from periodic insertion of pentagons and heptagons in the perfect hexagonal carbon network (see Fig. 9.1). In this sense, pentagons and heptagons in carbon nanocoils are essential supplies (not defects) in contrast to the cases of carbon nanotubes.[a] Structural stability of the proposed atomistic model has been examined by numerical simulations; the cohesive energies turned out to be $\sim$7.4 eV/atom, which is slightly higher than that of C_{60} (7.29 eV/atom) [270, 272].

Microscope measurements revealed that Young's modulus of carbon nanocoils is comparable to those of hexagonal graphene sheets [273], that is, several hundred GPa. It has been also found that carbon nanocoils can be elastically extended up to about three

[a]It is interesting to note that curvature and topology of carbon nanocoils lead to an enhancement of molecular hydrogen absorption on their external surfaces [271]. The enhanced storage originates from very high surface area and a large number of topological defects like pentagons and heptagons.

time versus the original coil length [227]. During such super-elastic elongation, interestingly, the average bond length between the nearest neighboring carbon atoms is almost invariant. Instead, flexibility in bond angles enables a significant change in the pitch and tubular diameter of the coils, resulting in the superelastic deformation of the whole system (see Section 9.5). These facts evidence that carbon nanocoils have good mechanical strength and flexibility. Against this background, potential applications for nano-sized mechanical devices have been strongly expected, though quantitative determination of their elastic (and non-elastic) properties and performance in actual applications remain largely to be examined.

9.2 Synthesis of Carbon Nanocoils

Experimental efforts in synthesizing carbon nanocoils of high quality have been triggered by the first transmission electron microscopy (TEM) observation of multi-walled carbon nanocoils in 1994 [274, 275]. Catalytic decomposition of acetylene was employed to fabricate carbon nanocoils having 30 nm in pitch and 18 nm in tubular diameter as shown in Fig. 9.2. The structural analysis based on electron diffraction method indicated that these coils are multi-walled, hollow, and polygonized such that they consist of short straight segments. These features agree with the theoretical prediction [270] that the coiling can result from the introduction of pentagon-heptagon pairs at regular distances in a straight carbon nanotube.

Since the first synthesis, many researchers have tried to make up these materials. Production of carbon nanocoils by chemical vapor deposition [276–280], laser evaporation [281], and opposed flow flame combustion method [282] has been reported to date. Recently the diameter control of carbon nanocoils via tuning the particle size of the nanoscale catalysts was suggested [283]. In addition to the multi-walled nanocoils with tubular diameters of 15–100 nm [277–280, 284], evidence of ultrathin single-walled carbon nanocoils (with both tubular diameter and pitch length down to 1 nm) was

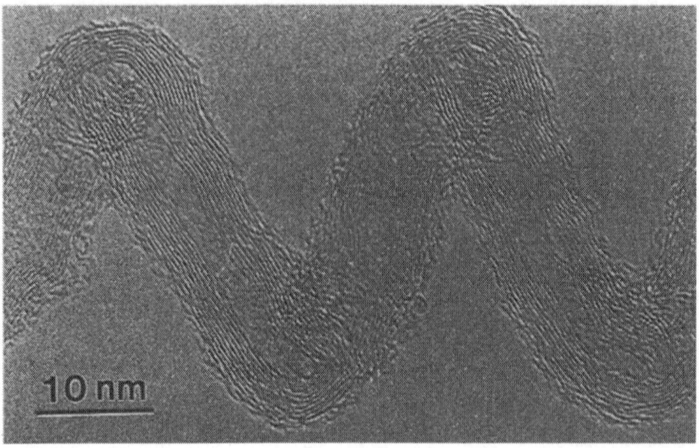

Figure 9.2 High-resolution TEM image of a well-graphitized carbon nanocoils with radius of about 18 nm, pitch about 30 nm, consisting of 10 graphitic tubes. Diameter of the innermost tube is about 2.5 nm. Reprinted from *Chem. Phys. Lett.*, **223**(4), Ivanov, V., Nagy, J. B., Lambin, Ph., Lucas, A., Zhang, X. B., Zhang, X. F., Bernaerts, D., Van Tendeloo, G., Amelinckx, S., and Van Landuyt, J., The study of carbon nanotubules produced by catalytic method, 329–335, Copyright (1994), with permission from Elsevier.

achieved by scanning tunneling microscopy (STM)[a] [285]. Readers can see Ref. [286] for a critical review on the synthesis of carbon nanocoils.

9.3 Microscopic Model of Carbon Nanocoils

A simple way to construct atomistic models of single-walled carbon nanocoils is based on periodic insertion of heptagons and pentagons in the predominantly hexagonal carbon network [287]. It starts from a piece of straight carbon nanotube depicted in Fig. 9.3(a). We suppose that in two opposite sides of the nanotube piece, one pair

[a] A scanning tunneling microscope is an instrument for imaging surfaces at the atomic level. It is based on the concept of quantum tunneling; when a conducting tip is brought very close to the surface to be examined, voltage difference between the two can allow electrons to tunnel quantum mechanically through the vacuum between them. The resulting tunneling current gives information about the atomic configuration over the surface considered.

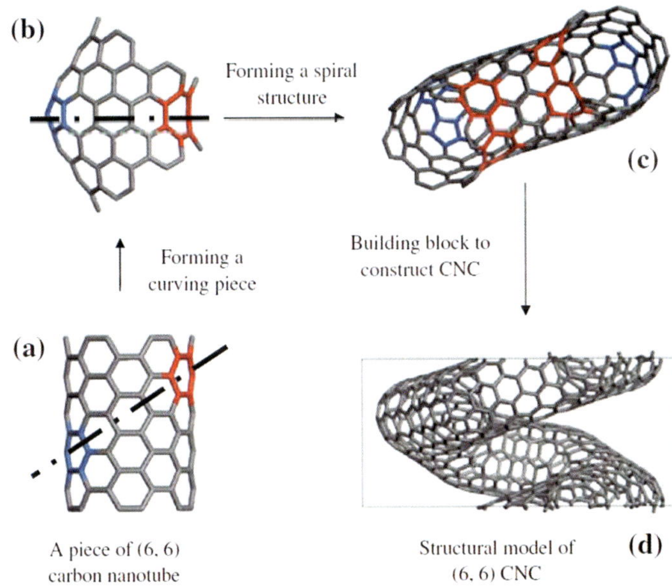

Figure 9.3 Procedures of constructing a carbon nanocoil of a (6,6) type from pieces of a (6,6) carbon nanotube. Reprinted with permission from Liu, L. Z., Gao, H. L., Zhao, J. J., and Lu, J. P. (2010). Superelasticity of carbon nanocoils from atomistic quantum simulations, *Nanoscale Res. Lett.*, **5**, 478–483.

of pentagons and one pair of heptagons are individually introduced at blue and red positions, respectively, marked in Fig. 9.3(a). Upon relaxation, the nanotube segment is bent around the non-hexagonal rings in order to release the strain energy [270, 288]. Introducing pentagons forms a cone shape, while introducing heptagons results in a saddle shape as shown in Fig. 9.3(b).

After preparing a number of segments, we connect one by one with a certain rotating angle to make a combined structure spiral (see Fig. 9.3(c)) that serves as a building block. Repetitive operation of this connecting procedure results in a seamless, complete nanocoil as demonstrated in Fig. 9.3(d). The index (n, n) in the plot means that the nanocoil is constructed from many pieces of the straight (n, n) nanotube. By changing the tube length at the two ends of the building block segment, or by varying the nanotube diameter,

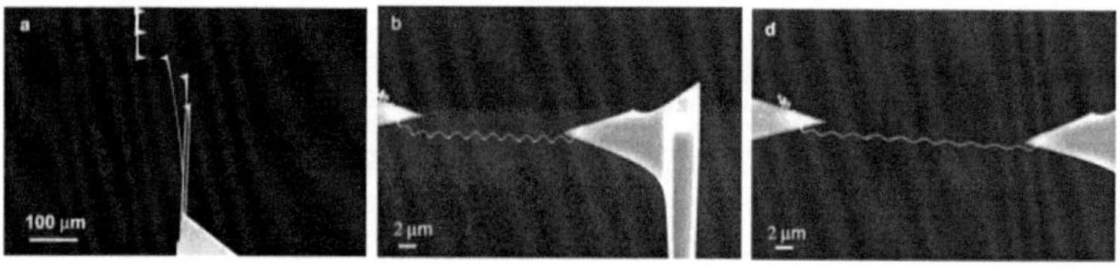

Figure 9.4 (a) Nanocoil clamped between two AFM cantilevers. The left cantilever is stiffer than the right cantilever. (b) Relaxed nanocoil prior to loading. (d) Nanocoil at a relative elongation of 33%. Reprinted with permission from Chen, X., Zhang, S., Dikin, D. A., Ding, W., Ruoff, R. S., Pan, L., and Nakayama, Y., Mechanics of a carbon nanocoil, *Nano Lett.*, **3**, pp. 1299–1304, 2003. Copyright (2003) American Chemical Society.

we can control coil diameter, coil pitch, and tubular diameter of a carbon nanocoil.

9.4 Spring Constant of Carbon Nanocoils

Spring-like behavior under loading is an important consequence of the helical structures inherent to carbon nanocoils. A direct tensile-loading test of determining their spring constants was done on multi-walled carbon nanocoils with outer tubular diameter of ~126 nm [264]. The experimental setup is shown in Fig. 9.4 A nanocoil is clamped between two opposed atomic force microscope (AFM) cantilevers, in which the left cantilever is the stiffer and the right one is the more compliant. After reaching the maximum 33% relative elongation, the load was slowly released, and the nanocoil relaxed to its original length and geometry. Remarkably, no plastic deformation was identified after the nanocoil was unloaded, which indicates the high elasticity of the nanocoils.

The spring constant of the nanocoil, K, is defined by the total applied load divided by the total elongation as same as macroscopic coils. Experimental date are shown in Fig. 9.5 [264]. At low-strain levels, the nanocoil behaves as an elastic spring with a spring constant $K = 0.12$ N/m, with a characteristic upturn in K at higher strain. Instead of elongation, the axial compression was also employed to deduce the stiffness of nanocoils as $K = 0.7$ N/m

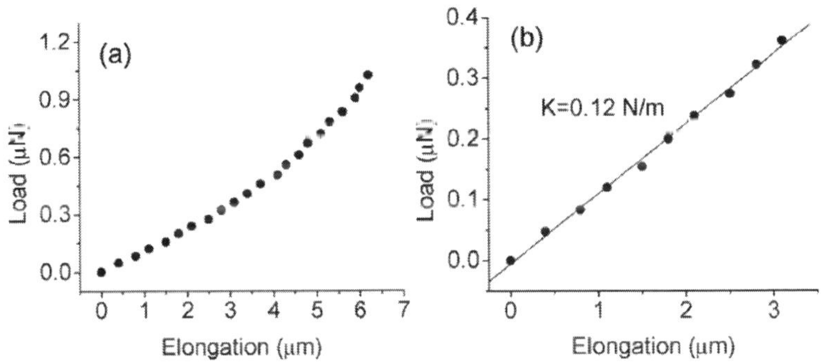

Figure 9.5 (a) Load versus elongation response of the nanocoil with outer tubular diameter ~126 nm. (b) Low-strain region response. Reprinted with permission from Chen, X., Zhang, S., Dikin, D. A., Ding, W., Ruoff, R. S., Pan, L., and Nakayama, Y., Mechanics of a carbon nanocoil, *Nano Lett.* **3**, pp. 1299–1304, 2003. Copyright (2003) American Chemical Society.

[265]. It is noteworthy that the spring constant obtained above is much smaller than those of other nanomaterials having coiled geometry: the superlattice ZnO nanohelices (~4 N/m) [266], metal-coated Si nanocoils (= 8.75 N/m) [267], and an array of tightly packed SiO nanocoils (= 590 N/m) [268] are by far stiffer than carbon nanocoils.

9.5 Superelasticity of Carbon Nanocoils

In Section 9.4, we have seen that carbon nanocoils show elastic response up to, at least, a relative elongation of ~33%. It naturally poses a question as to what are the elastic limits under elongation and compression of carbon nanocoils. These limits have been numerically examined using tight-binding calculations [264]. Surprisingly, the results suggested that the nanocoils accept elongation (compressive) strains up to about 60% (20–35%). Above such elastic limits, the carbon nanocoils will undergo plastic deformation. The above computational results show superior superelasticity in carbon nanocoils. The superelasticity of carbon nanocoils may lead to some applications in nanoscale materials and devices, for

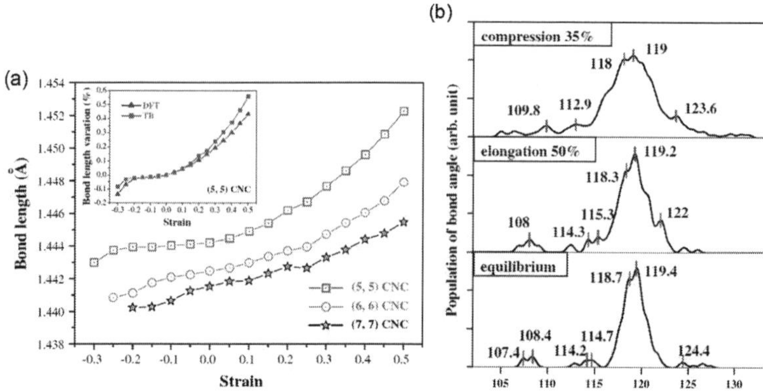

Figure 9.6 (a) Variation of average C–C bond length in carbon nanocoils under elongation (positive strain) and compression (negative). The inset shows the percentages of bond length variation with regard to the equilibrium state for a (5, 5) carbon nanocoil obtained by density-functional-theory (DFT) based and tight-binding (TB) based calculations. (b) Bond angle distribution of (5, 5) carbon nanocoil under large elongation (50%) and compressive strain (35%), compared to the equilibrium case. Reprinted with permission from Liu, L. Z., Gao, H. L., Zhao, J. J., and Lu, J. P. (2010). Superelasticity of carbon nanocoils from atomistic quantum simulations, *Nanoscale Res. Lett.*, **5**, 478–483.

example, shape memory, elastic energy storage, buffer, nano-spring in nanoelectromechanical system (NEMS), and so on.

It should be emphasized that within the elastic strain ranges considered, the topology of carbon atomic structure is still retained, with an average bond length only increased by less than 1%. Figure 9.6(a) demonstrates the changes in average C–C bond lengths of carbon nanocoils during elongation and compression. The average bond length is shown to be very robust under external strains of both directions. With elongation strain up to 50%, the increase in average bond length is less than 1% for all samples. Moreover, the average bond length is very slightly reduced under compression. This phenomenon can be understood by the substantial strength of sp^2 chemical bonds. In order to avoid significant changes of C–C bond lengths, the relative orientations of neighboring C–C bonds (i.e., bond angles) alter during compression or elongation. Figure 9.6(b) shows the full width at half maximum (FWHM) of bond angle distribution for a (5, 5) carbon nanocoil. Clearly the FWHM increases during elongation or compression, justifying the scenario mentioned above.

With increasing tubular diameter, the variation of average bond length in the nanocoil is less sensitive to elongation strain (see Fig. 9.6(a)), implying that the nanocoil can undertake higher strain. On the contrary, the elastic limit of compression for a carbon nanocoil reduces with increasing tubular diameter. For example, the maximum compressive strain is 35% for (5, 5) carbon nanocoil and 20% for (7, 7) carbon nanocoil.

9.6 Coil Geometry Statistics

Toward the mass production of carbon nanocoils, it is crucial to understand the relation between the geometric structure and synthesis methods. In fact, a wide variety of synthesis techniques for carbon nanocoils has been reported thus far as mentioned in Section 9.2. Hence, the possibility of controlling the coiled geometry should be examined in order to accomplish the mass production. The need has motivated a statistical analysis of the characteristic parameters, coil diameters and pitch (see Fig. 9.7), of coils produced under various experimental conditions [289].

The plot in Fig. 9.7 shows the statistical distribution of carbon nanocoils grown by chemical vapor deposition. A three-dimensional plot was used to analyze the number of coils exhibiting a certain pair of coil diameter and pitch value. It should be borne in mind that the data come from three different laboratories and from a number of at least 15 experimental cycles. Hence, there can be no correlation in coiled geometry between samples generated by different experiments. Nevertheless, the plot clearly shows the presence of specific regions in which the number of coils produced is significantly higher than in the neighboring regions. The most preferred configuration is delineated by the crossing of the region of 50–70 nm pitch and 20–60 nm diameter. In addition, a slightly less pronounced region is found between pitch 30 and 50 nm and the same diameters.

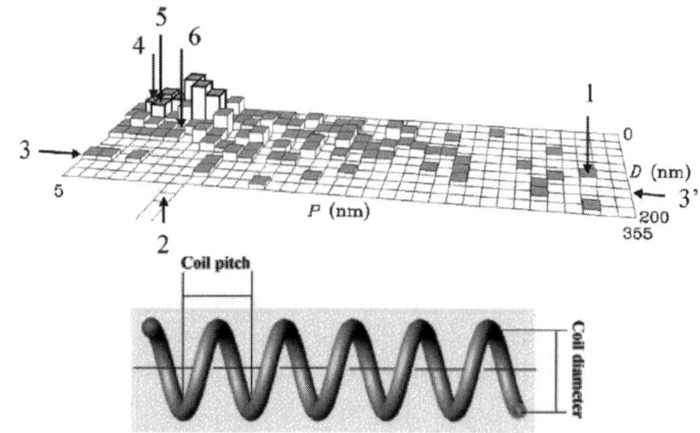

Figure 9.7 Top: Three-dimensional plot of number of coils vs. coil diameter (D) and pitch (P) for coiled nanotubes grown by chemical vapor deposition. The two primary stability islands are highlighted in bold lines. Reprinted from Carbon, **43**(8), Szabó, A., Fonseca, A., Nagy, J. B., Lambin, Ph., and Biró, L. P., Structural origin of coiling in coiled carbon nanotubes, 1628–1633, Copyright (2005), with permission from Elsevier. Bottom: A schematic drawing of a carbon nanocoil showing the coil diameter and coil pitch. Reprinted from *Carbon*, **42**(4), Lu, M., Liu, W. M., Guo, X. Y., and Li, H. L., Coiled carbon nanotubes growth via reduced-pressure catalytic chemical vapor deposition, 805–811, Copyright (2004), with permission from Elsevier.

It should be noted that a fraction of 24.5% of the total number of coils is found in the two stability regions; this has to be compared with 2% that would correspond to the same area if a uniform distribution of coils over the entire area of the plot would be found. The analysis of the coil diameter distribution of Fig. 9.7 shows that half of the coils are found in the range of 50–70 nm, while again, half of the coils have a pitch in the range of 30–80 nm. Since the analysis is done for an ensemble of nanocoils produced by different experiments, the possibility of accidental coincidences should be ruled out. Therefore, these data can be considered an indirect proof that the way in which carbon nanotubes are coiled has an intrinsic, structural origin and it is not decided by particular conditions under which coiling occurs in actual experiments.

9.7 Cushioning Effect

A striking implication of high resilience of carbon nanocoils is their potential use as "cushion" against impact. An ensemble of vertically aligned nanocoils arranged in bundles, which is called a carbon nanocoil forest, may serve as a shock-wave mitigation materials and an energy-absorbing device. To this end, mechanical response of a forest against high-strain-rate deformation has been studied using a simple and conventional experimental approach [291].

Figure 9.8 shows the setup for measuring the mechanical response consists of impacting ball and the array of nanocoils. The thickness and density of the forest are 100 μm and 100 μm^{-2},

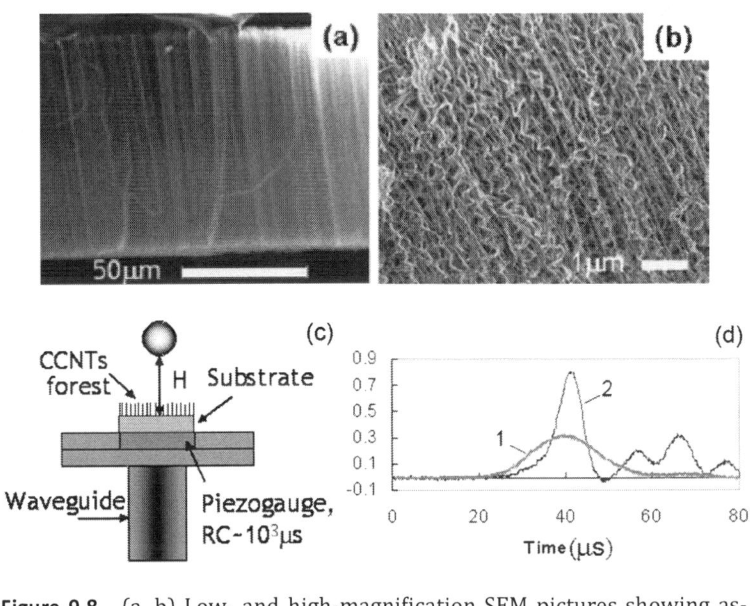

Figure 9.8 (a, b) Low- and high-magnification SEM pictures showing as-grown forest of carbon nanocoils. (c) Experimental setup used for testing the forest of carbon nanocoils. (d) Experimentally determined force vs. time response obtained for the nanocoil forest (curve 1) and the bare quartz substrate (curve 2) when impacted with a 2 mm diameter steel bead (0.02 g) dropped from a height of 2 mm. Reprinted with permission from Daraio, C., Nesterenko, V. F., Jin, S., Wang, W., and Rao, A. M., Impact response by a foamlike forest of coiled carbon nanotubes, *J. Appl. Phys.*, **100**, 064309-1–4, (2004). Copyright 2004, American Institute of Physics.

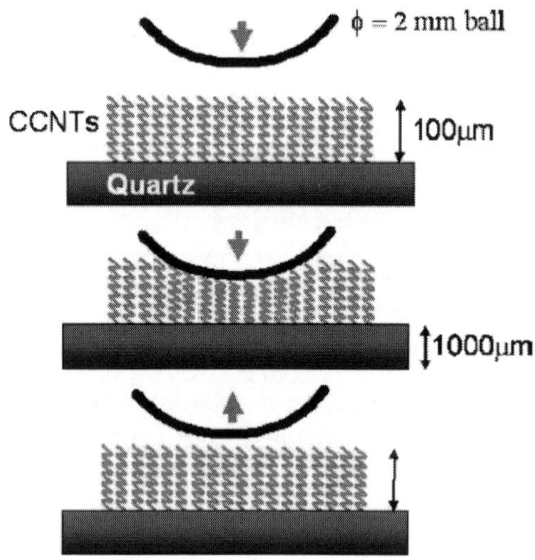

Figure 9.9 Schematic diagram showing the stages of interaction and full recovery of the carbon nanocoil array during and after the impact. Reprinted with permission from Daraio, C., Nesterenko, V. F., Jin, S., Wang, W., and Rao, A. M., Impact response by a foamlike forest of coiled carbon nanotubes, *J. Appl. Phys.*, 100, 064309-1–4, (2004). Copyright 2004, American Institute of Physics.

respectively, in which the nanocoils are arranged in bundles due to van der Waals force attraction between them. The single coiled nanotubes in the forest had a narrow diameter distribution around 20 nm with a coiling pitch of ~500 nm (Fig. 9.8). Against the forest, high-strain-rate impacts were generated by dropping a stainless steel bead (2 mm diameter, mass of 0.03 g) from various heights at room temperature. The calculated velocities of the impacts varied between 0.2 and 2.0 m/s.

The typical contact force-time response after a small amplitude impact (0.2 m/s striker velocity) on the surface of the film is shown in Fig. 9.8(d), curve 1. Repeated experiments demonstrate identical behavior of contact force. For comparison, the contact response on the bare substrate (without the forest) under identical impact conditions was also measured and presented in Fig. 9.8(d), curve 2. The presence of the forest dramatically changes the slope of the contact force and attenuate the amplitude of the pulse. A perfectly elastic response is also noticeable from the symmetry of curve 1. Such a resilient system could find applications in micro/nanoelectromechanical systems and actuators as well as coating for protection purposes.

It is noteworthy that that the deformation of this forest under vertical impacts exhibits a strongly nonlinear contact interaction law [292]. This means that the nanocoils respond to dynamic loading as perfect elastic nonlinear springs that fully recover their original lengths under the impact conditions. The nonlinear behavior is fully described by considering the entanglement of adjacent nanocoils in the superior part of the forest surface [292]. This entanglement among neighbors is due to the bending of the coil tips produced by the ball impact.

Chapter 10

Irradiation-Based Tailoring

It is never too late to become what you might have been.
— George Eliot

10.1 What Can Be Done by Irradiation?

High-energy beam irradiation can have beneficial effects in manipulating the carbon nanotube morphology, especially when combined with heat treatment. The manipulation is achieved through the irradiation-induced displacement of carbon atoms from their equilibrium positions. When the kinetic energy transferred from incident particles to the carbon nuclei is larger than the displace threshold, a carbon atom leaves its initial position and is pushed far away from the nanotube or is quenched into a metastable defect configuration[a] (see Fig. 10.1, for example). Hence, bombardment with energetic particles can be used for the sputtering of atoms

[a]Irradiation-induced defects in carbon nanotubes can be detected by various experimental techniques; they include microscope scanning [293, 294], optical (Raman) scattering [295–297], and X-ray photoelectron spectroscopy[298–300]. In the latter two measurements, the intensity and width of specific scattering peaks are sensitive to the type of carbon bondings.

Elastic and Plastic Deformation of Carbon Nanotubes
Hiroyuki Shima and Motohiro Sato

ISBN 978-981-4310-75-8 (Hardcover), 978-981-4364-15-7 (eBook)
www.panstanford.com

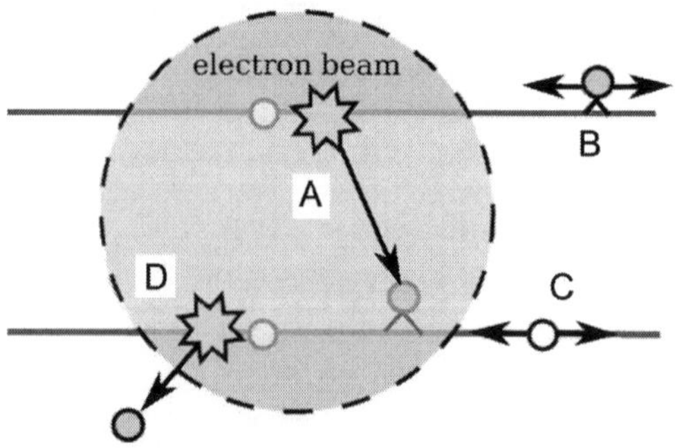

Figure 10.1 Schematic of the most relevant processes in carbon nanotube tailoring by electron-beam irradiation. (a) Electron impact creating a defect pair by displacing a carbon atom; the white sphere represents a vacancy and the gray sphere an adatom. (b, c) Adatom and vacancy migration. (d) Creation of a vacancy by sputtering the displaced atom. Reprinted with permission from Krasheninnikov, A. V., and Nordlund K., Ion and electron irradiation-induced effects in nanostructured materials, *J. Appl. Phys.*, **107**(7), 071301, 2010. Copyright 2010, American Institute of Physics.

in pre-defined areas, and successive sputtering treatment makes it possible to alter the local (or larger-scale) geometry of nanotubes.

Irradiation-based tailoring with high-energetic beams has two main advantages. The first is the controllability of the area or volume that is subjected to attack by irradiation. Ion beams, for example, can be focused onto spots of a few nanometers in diameter. Much higher resolution can be achieved using electrons emitted from field emission guns, which offer the focusing as small as 1 Å in diameter that is less than the C–C bond length in nanotubes. Besides, particle beams can be scanned faster than the ordinary mechanical-tip-based techniques, allowing the treatment of a large portion of structures within a reasonable time.

The second advantage is the self-repairing property of carbon nanotubes as the response to the particle beam irradiation. It was experimentally found that irradiation-induced damage (i.e., removal of carbon atoms) can be annealed spontaneously through dangling bond saturation[a] [302–305]. As a result, severe radiation

[a] This self-organized reconstruction is universal to sp^2-hybridized graphitic structures [301].

damage that could destroy the original tubular structure can be minimized. These facts imply the possibility of a desired modification of nanotube geometry by introducing artificial reconstruction of graphitic networks. In fact, many attempts of tailoring the nanotube's structure with high-energetic beams were demonstrated to date [27, 219, 306–312], as we will see in Sections 10.5, 10.6, and 10.8.

10.2 High-Energetic Particles Workable

Thus far, nanotube irradiations with different kinds of energetic particles have been reported, such as γ-rays [313, 314], electrons [27, 315, 318, 319], protons [316, 317, 320], and ions [296, 309, 321, 322]. These particles possess the ability to alter the mechanical (and other physical) properties of the nanotubes; a strong stiffening of nanotube bundles obtained after electron irradiation [27] and a dramatic increase in the nanotube's electrical resistivity after Ar^+ ion bombardment [309] are only a few to mention. It was also found that [323] γ-radiation significantly affects the mechanics of carbon nanotube composites, i.e., mixture solids of a certain amount of nanotubes with an embedding host matrix such as polymers, ceramics, or metals; see Chapter 13. Such knowledge of irradiation effects on nanotube composites will aid design for radiation-resistant materials, e.g., against nuclear radiation.

Among the choices, accelerated electrons provide three main advantages compared with other energetic particles. First, both the rate of energy transfer and the irradiation dose can be so low that only individual defects are generated. Second, the energy of the incident electron can be finely tuned around the threshold just above which defects are created. Finally, electron beams can be focused to form a nanometric probe, allowing control of the irradiated area with high spatial resolution [324]. These three merits are the reason why we focus on the electric beam irradiation in the remainder of this chapter.

10.3 Self-Healing of Vacancies

10.3.1 *Overview*

In view of irradiation-based tailoring, vacancies are the most prolific defects in nanotubes. A vacancy originates from missing atoms in the ideal crystalline lattice. In usual solids, it keeps its void structure without curing as time passes; however, it is not the case in carbon nanotubes. The removal of carbon atoms from the purely hexagonal lattice leads to a local reconstruction so as to maintain its coherent network structure with cylindrical geometry. For this reason, carbon nanotubes are often referred to as "*self-healing*" materials [28].[a,b]

Figure 10.2 illustrates a time evolution of a reconstructing SWNT; the simulation starts a nearly perfect carbon nanotube (see Fig. 10.2(a)) from which carbon atoms are extracted homogeneously step by step. During the course of the simulation, most vacancies try to recombine, thus forming a highly defective carbon cylindrical network. Figure 10.2(b) presents the reconstructed rough surface of the nanotube when half of the atoms in the original model were extracted. Fifteen pairs of 5–7 defects are present in this final topology,[c] playing the primary role in maintaining the cylindrical shape of the defective tube. The tube's diameter decreases from 1.36 nm (Fig. 10.2(a)) to an average value of 0.7 nm (Fig. 10.2(b)) due to loss of carbon atoms. If carbon extraction is performed within a localized region, the defective surface is not able to reconstruct in a disordered sp^2 network but rather creates atomic linear chains of

---

[a] Similar self-healing behavior was seen in flat graphetic sheets [253, 327, 328]. But in nanotubes, the curvature around the tubular axis causes the reconstruction to be much stronger, as we will see in Section 10.4.

[b] Self-healing of irradiation-induced defects (vacancies and associated dangling bonds) was evidenced in telescopic motion of defective MWNTs [28]. Defects that were intentionally introduced by electron irradiation led to an increase in friction, but self-healing rapidly optimized the atomic structure and restored smooth motion of the sliding shells.

[c] Non-hexagonal rings like squares, pentagons, heptagons, octagons, nonagons, and decagons were observed at certain stages of the surface reconstruction. However, the unstable high-membered rings are found to disappear by Stone–Wales mechanism (see Chapter 6), thus leading the structure to be mainly constituted of five-, six-, and seven-membered rings.

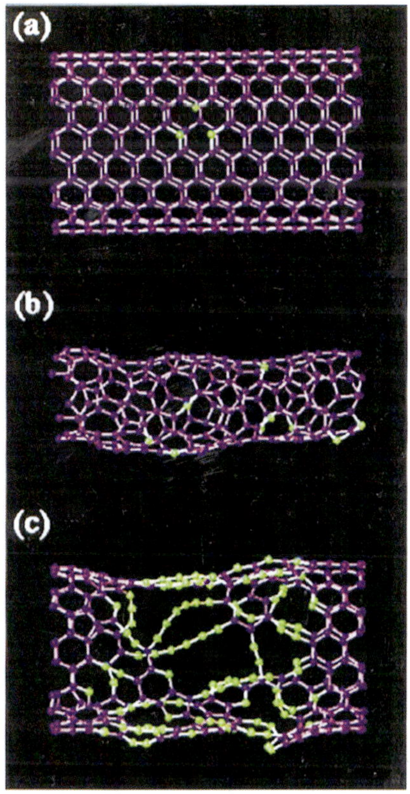

Figure 10.2 (a) The atomic network of a (10,10) SWNT with diameter of 1.36 nm after one-atom extraction. (b) Surface reconstruction of the nanotube after a random extraction of 200 carbon atoms. Although the reconstructed surface is highly detected, the carbon system is still a rough cylinder with diameter value averaged around 0.7 nm. Although each vacancy creation produces three two-coordinate carbon atoms (in green), their number is very small after the reconstruction, illustrating its stability as a nearly pure sp^2 disordered carbon network. (c) Surface reconstruction of the nanotube after an extraction of 80 carbon atoms in a localized region of the cylinder (center of the image). Linear atomic chains of carbon are bridging the two opposite undefected parts of the nanotube. Reprinted with permission from Ajayan, P. M., Ravikumar, V., and Charlier, J.-C., *Phys. Rev. Lett.*, **81**, 1437–1440, 1998. Copyright (1998) by the American Physical Society.

carbon (Fig. 10.2(c)), which is called "carbynes".[a,b] In either case, the persistence of cylindrical network structure is clearly observed even after a significant amount of atoms is lost.

10.3.2 *Microscopic view*

What happens if a few-atomic-scale vacancy is placed in isolation on the hexagonal wall? In general, a one-atom vacancy results in one pentagon ring and a dangling bond,[c] while a two-atom vacancy leads to two pentagons and an octagon without any dangling bonds. In either event, the resulting cured network contains locally non-hexagonal atomic rings.

Such reconstruction procedure near vacancies is shown in Fig. 10.3. The Fig. 10.3(a) tells us that there are two choices of reconstruction of a single vacancy (SV) in an armchair carbon nanotube. Namely, the original unstable configuration (top panel) transforms into stable or metastable states by saturating two of the three dangling bonds around the vacancy. The latter two states are distinguished by the orientation of the "pentagon bond" that newly emerges near the missing atom. The final state should be metastable if the bond is parallel to the tube axis, or stable when the bond is almost perpendicular to the tube axis. The relative energetic stability of the two states will be discussed in more detail in Section 10.4.

Figure 10.3(b) depicts a reconstruction associated with three different kinds of small vacancies. A divacancy (DV) transforms to an agglomeration of five- and eight-membered rings, where the orientation of the newly emerging pentagon bonds is uniquely determined (i.e., perpendicular to the missing C–C bond) in contrast with the panel (a). Figure 10.3(b) also shows the possibility that a pair of one SV and one adatom forms a metastable 5-7-7-5

[a] Carbyne is a linear monoatomic chain with *sp* orbital hybridization. It is called by another name as "linear acetylenic carbon" [325]. The carbyne is of considerable interest to nanotechnology as its Young's modulus is 40 times higher than that of diamond [326].

[b] Earlier simulations have shown that similar atomic chains can occur if a nanotube is pulled in tension until its thickness in the necked region becomes on the other of one atomic scale [260].

[c] It is important that unlike DVs, SVs always have a dangling bond, so that a perfect reconstruction around a SV is not possible.

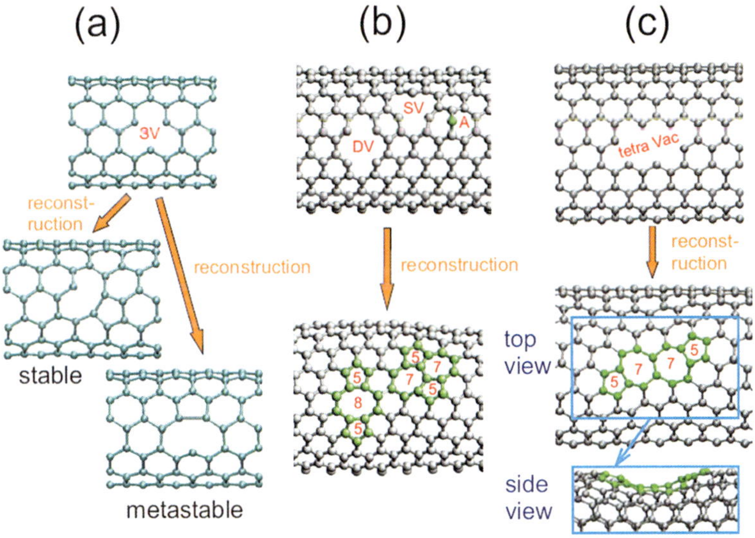

Figure 10.3 Reconstructions of the atomic network near vacancies. (a) A single vacancy (SV) state tends toward stable or metastable configurations depending on the orientation of the new weak "pentagon bond" (see text). (b) A divacancy (DV) transforms to an agglomeration of five- and eight-membered rings. A SV and an adatom may form a metastable 5-7-7-5 defect. (c) Reconstruction of a tetravacancy. The effective size of the "hole" decreases at the expense of local diameter reduction. Reprinted with permission from Krasheninnikov, A. V., and Nordlund K., Ion and electron irradiation-induced effects in nanostructured materials, *J. Appl. Phys.*, **107**(7), 071301, 2010. Copyright 2010, American Institute of Physics.

defect, which may occur depending on the relative position of the neighboring SV and adatom.

Figure 10.3(c) shows reconstruction of a tetravacancy. The salient feature is that the effective size of the "hole" decreases (see top view) at the expense of local diameter reduction (side view). This vacancy-induced reduction in local diameter implies that if many large-scale vacancies are dispersed intentionally, then we will observe a global diameter reduction over the whole tube surface damaged (as seen in Fig. 10.2(b)). See Sections 10.5 and 10.6 for the experimental evidence of such global shrinkage.

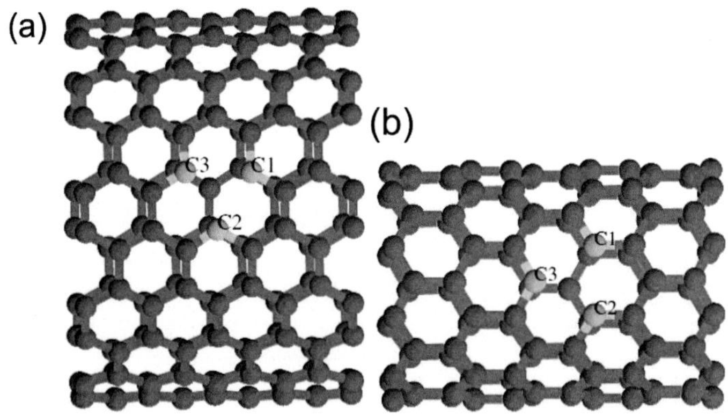

Figure 10.4 Configurations of ideal vacancies in: (a) (n, n) tubes and, (b) $(n, 0)$ tubes. Carbon atoms in the vicinity of the defect are labeled as C1, C2, and C3, with the light shading. Reprinted with permission from Lu, A. J., and Pan, B. C., *Phys. Rev. Lett.*, **92**, 105504-1–4, 2004. Copyright (2004) by the American Physical Society.

10.4 Relax from a One-Atom Vacancy

10.4.1 *In Armchair SWNT*

As shown in Fig. 10.3, the atomic structure in the vicinity of an isolated vacancy transforms into more energetically favored configuration through dangling bond saturations. It has been theoretically suggested that the energetic stability of the resulting configurations depend on the radius and chirality of the tube. This section outlines what relaxed structure is obtained around a single vacancy (SV) for a given tube radius and chirality.

We first address the case of an SW in armchair SWNT indexed by (n, n). We learned in Section 10.3.2 that a SV in armchair SWNTs is unstable; two of the three dangling bonds prefer to recombine each other, forming a pentagon ring plus one dangling bond remaining. Figure 10.4(a) visualizes an example of the (10,10) nanotube; there are three dangling bonds around the unrelaxed vacancy. Through a relaxation, the system will achieve an optimized structure in which the two carbon atoms, labeled by C1 and C2 in Fig. 10.4(a) (or C1 and C3, equivalent symmetrically) form a weak C–C bond. This new C–C bond tilts with respect to the tube axis. Thus, the local

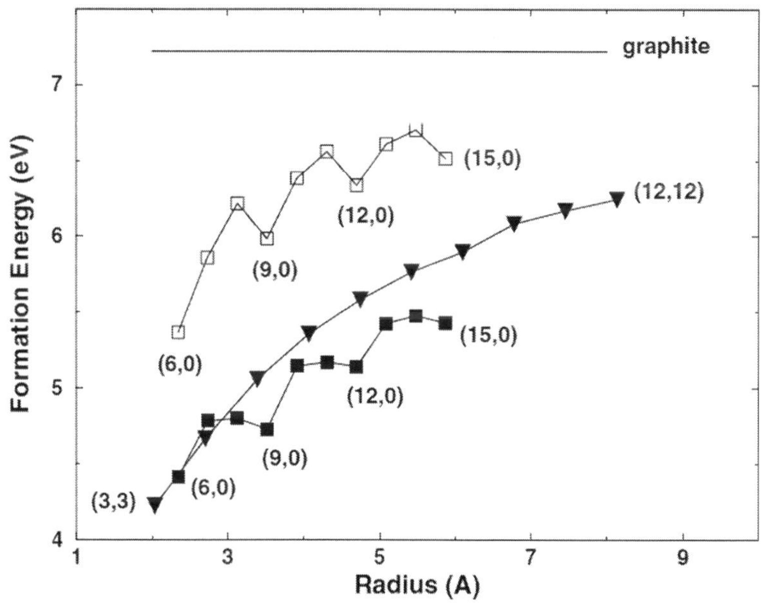

Figure 10.5 Defect formation energies as a function of radius of a tube. Down triangles stand for 5-1DB defects in (n, n) tubes, filled squares for 5-1DB-P defects in $(n, 0)$ tubes, and open squares for 5-1DB-T defects in $(n, 0)$ tubes. Reprinted with permission from Lu, A. J., and Pan, B. C., *Phys. Rev. Lett.*, **92**, 105504-1–4, 2004. Copyright (2004) by the American Physical Society.

atomic network around the point defect features one pentagon and one dangling bond in C3 [252], which is called a 5-1DB defect [303]. The scenario essentially holds for the other defected (n, n) tubes. It is important to mention that another choice of the bond formation between C1 and C3, parallel to the tube axis, is energetically unfavored; the preference of the tilted bond (C1–C2 or C2–C3) originates from the fact that the curvature of the atomic network around the tube axis breaks the trigonometric symmetry of a flat graphetic sheet [304, 305].

Figure 10.5 plots the formation energy curves of 5-1DB (solid triangle) in various armchair nanotubes. The formation energy E_{form} measures the energetic cost to make the relaxed configuration from an M-atom vacancy state, and is defined by

$$E_{\text{form}} = E_{\text{def}} - E_{\text{perf}} \times \frac{N_{\text{perf}} - M}{N_{\text{perf}}}. \quad (10.1)$$

Here, E_{def} and E_{perf} are the total energies of the defective (fully relaxed) and perfect tubes, respectively, and N_{perf} is the number of atoms in the nanotube before the vacancy is created. The plot clearly shows that the defect formation energies for (n, n) tubes increases monotonically with increasing tube radii. In addition, all energies shown in the energy curve are lower than the formation energy of a single vacancy in graphite. This is because the curvature may give rise to a shorter bond in the non-hexagonal carbon ring rather than the bond in a flat graphite sheet. This shortness of the bond results in a considerable drop in the formation energy of these non-hexagonal rings. The formation energy decreases when the diameter becomes smaller, which can be understood in terms of the easier reconstructions of the nanotube atomic network.

10.4.2 *In Zigzag SWNT*

Unlike the armchair cases, some care must be taken for the nanotubes with zigzag symmetry characterized by the index $(n, 0)$. For the $(n, 0)$ series, the stable (relaxed) configurations are dependent on n, i.e., on the tube radius. The local structures around the missing atom (Fig. 10.4(b)) do convert to a relaxed pattern in which the formed new bonds, C1–C2, are perpendicular to the tube axes; this defect structure is named as 5-1DB-P. On the other hand, in the cases of $(n, 0)$ with $n > 9$, we obtain the bond formation of C1-C3 or C2-C3, whose orientations tilt about the tube axes, named as 5-1DB-T. Nevertheless, the tendency to form the 5-1DB-T state appears controversial when we calculate the formation energy for $(n, 0)$ tubes. Figure 10.5 clearly tells us that E(5-1DB-T) $>$ E(5-1DB) $>$ E(5-1DB-P) for given tube radius; hence, the 5-1DB-P defect is the most favored energetically among the three kinds of defects (except for the cases of very small radii).

The apparent controversy is resolved once we get aware that in defected $(n, 0)$ tubes with $n > 9$, the initial structures containing single vacancies cannot directly convert to the most favored configuration 5-1DB-P but to a metastable state 5-1DB-T. This immediately raises a question: How much energy is associated with defect transformation from 5-1DB-T to 5-1DB-P for these cases? In the case of the (10,0) tube, for example, the two stable states are separated by an energy barrier of ~1.9 eV or less. Hence, thermal agitation and/or bond stretching are needed to overcome the barrier to settle down the truly energy-minimum state (i.e., 5-1DB-P).

It is also noteworthy that the formation energy curves of 5-1DB-P and 5-1DB-T for $(n, 0)$ tubes exhibit steplike and sawtoothlike features, respectively. Interestingly, both sawtoothlike and steplike features are periodic, and such periodicity is somehow characterized by the lower formation energies of defected $(n, 0)$ tubes with $n = 3m$ $(m = 2, 3, 4, 5)$ compared with their neighbors, $(n - 1, 0)$ and $(n + 1, 0)$ tubes. Attempts to explain the periodicity has been made in Ref. [303], while the full understanding seems not to have been reached.

10.5 Radial Shrinkage of SWNTs

An important consequence of the vacancy reconstruction is shrinkage of the graphene cylinders due to the continuous loss of atoms [319]. The shrinkage implies the utility of the electron-beam irradiation for adjusting the tube's diameter by tuning the irradiation time and area. Figure 10.6 shows the result of an early experiment [252]. Uniform irradiation of an SWNT led to surface reconstruction and a drastic dimensional change, as a corollary of which the apparent tube diameter decreased from 1.4 to 0.4 nm, and finally the tube broke. The radius shrinkage originates from dangling bond saturations near the missing atoms created by energetic electrons.

Recent work in 2009 [329] examined the diameter-dependence of the SWNT stability against electron-beam erosion. It was experimentally found that SWNTs with small diameters degrade faster than their larger-diameter counterparts. To obtain the result in the measurement, both large (~2–3 nm) and small (~1 nm) diameter SWNTs were located next to each other as depicted in Fig. 10.7; this setup enables the same electron beam dosage to be applied to both narrow and large diameter SWNTs and provides

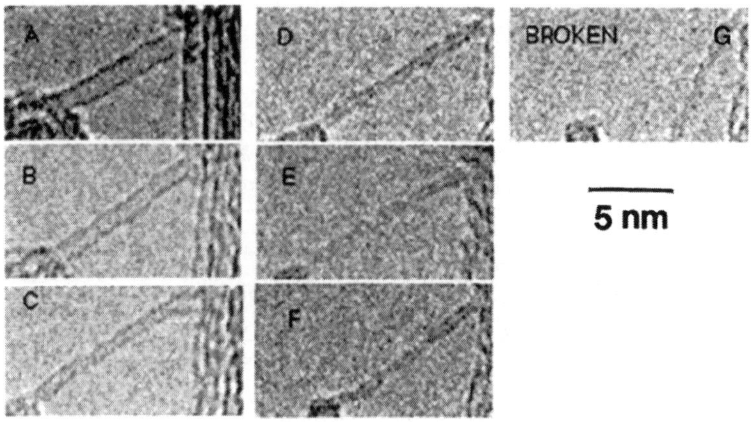

Figure 10.6 Controlled electron irradiation of an SWNT segment bridged across a hole in a carbon grid. The diameter of the original nanotube (a) is approximately 1.4 nm. The tube has shrunk drastically in diameter during the irradiation. The image in (f) shows the smallest diameter (~0.4 nm) that was visible before the tube broke (g). Reprinted with permission from Ajayan, P. M., Ravikumar, V., and Charlier, J.-C., *Phys. Rev. Lett.*, **81**, 1437–1440, 1998. Copyright (1998) by the American Physical Society.

a true comparison of diameter-dependent effects. Figure 10.7 shows a time evolution of three SWNTs with different diameters, one large and two small, with 20 seconds between each image. After 100 seconds of constant electron beam irradiation (see Fig. 10.7(f)), the two smaller SWNTs have undergone significant structural deformation, while the larger SWNT remains relatively unchanged.

The low stability of smaller-diameter SWNTs under irradiation is consistent with the argument in Section 10.4, as is attributed to the curvature-induced mechanical strain of rolled-up graphene sheet [330]. When a sheet of graphene is rolled up to form an SWNT, strain is induced on the carbon bonds due to the curvature of the SWNT. The narrower the diameter of the SWNT, the tighter the graphene sheet has to be rolled to form an SWNT and the higher the strain in the carbon bonds. As a consequence, higher strain in the chemical bonds results in reduced stability and increased reactivity [331].

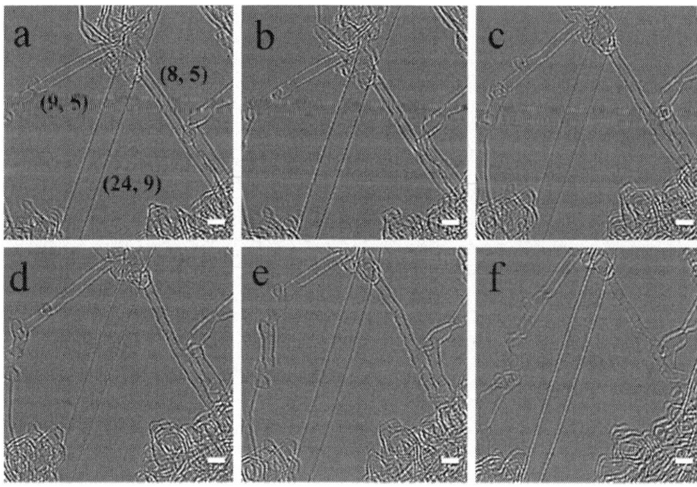

Figure 10.7 Time series of high-resolution TEM images showing the SWNTs with smaller diameter ((8,5) and (9,5)) and are destroyed faster than the larger-diameter SWNT (24,9). Scale bar indicates 2 nm. Reprinted with permission from Warner, J. H., Schäffel, F., Zhong, G., Rümmeli, M. H., Büchner, B., Robertson, J., and Briggs, G., Investigating the diameter-dependent stability of single-walled carbon nanotubes, *ACS Nano*, **3**, 1557–1563, 2009. Copyright (2009) American Chemical Society.

10.6 Internal Collapse of MWNTs

Similar to SWNTs, electron irradiation of MWNTs causes the formation of vacancies on their walls and eventual amorphization. Of particular interest is the high-dose irradiation of MWNTs at high temperature. Figure 10.8 shows the structural evolution of an irradiated MWNT at 900 K. With time, radial shrinkage of all shells is observed prior to the collapse of the innermost shell that starts at (b). It should be noticed that during the shrinkage, all shells (except for the collapsed innermost shell) remained temporarily intact without breakage or disintegration, though material was lost. As time passed, the innermost shell finally disintegrated (see Fig.10.8(b)) and collapsed into a morphology of a double cone (Fig.10.8(c)). During the collapse, an aggregation of irregular graphitic cages emerged in the core cavity just outside the irradiated area.

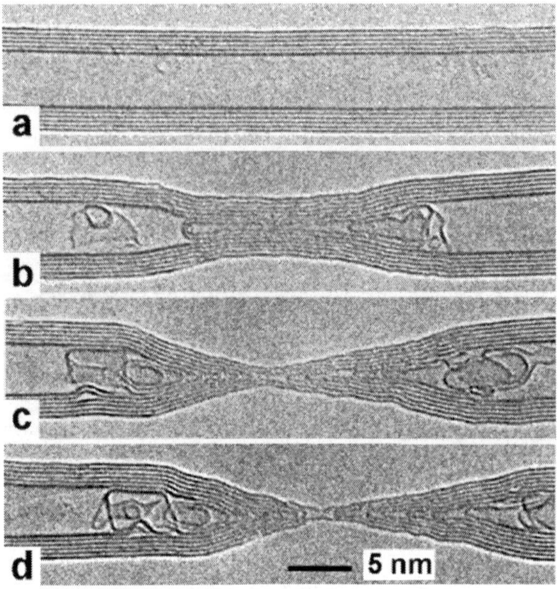

Figure 10.8 Morphological evolution of an MWNT under electron irradiation at 900 K. An electron beam with a diameter of 15 nm and a beam current density of $\sim$450 A/cm^2 was focused onto the central part of the tube. The irradiation time: (a) $t = 0$; (b) $t = 160$ s; (c) $t = 340$ s; (d) $t = 820$ s. Reprinted with permission from Banhart, F., Li, J. X., and Krasheninnikov, A. V., *Phys. Rev. B*, **71**, 241408(R)-1–4, 2005. Copyright (2005) by the American Physical Society.

The most important observation is the fact that the outer shells shrank but remained undamaged; it is always the inner shells that break in such a way that the two halves form cones with closed caps. During the successive creation of the cones associated with inner-wall breakings, two sets of the inner cones facing each other, which are both encapsulated the shrunken outer walls move away along the tube axis with the opposite directions; eventually, the last remaining outermost wall broke so that two separate multi-shell cones were left (not shown in Fig. 10.8). Such a behavior was explained in terms of a lower stability of the inner shells under irradiation due to a higher curvature of the atomic network [331] and fast diffusion of carbon interstitials through the inner hollow in the axial direction [332, 333].

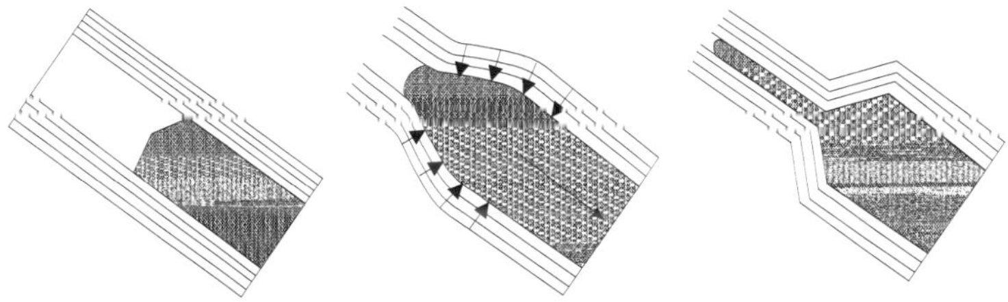

Figure 10.9 Simplified geometry of an irradiated MWNT partly filled with a Fe_3C crystalline nanowire (dark region) in the hollow cavity. Left: The MWNT before irradiation. Center and right: Irradiation-induced collapse of the inner walls that cause deformation of the Fe_3C crystal. Compressive forces (indicated by the small arrows) from the surrounding walls lead to a thinning of the encapsulated crystal and its sliding downward along the tube axis (large arrow). Adapted from Ref. [334].

10.7 Inner-Wall Corrugation by Outer-Wall Erosion

Irradiation-induced radial shrinkage of MWNTs was found to cause high pressure, higher than 40 GPa, that exerts on the innermost hollow cavity [334]. Figure 10.9 shows the evolution of an MWNT encapsulating a Fe_3C crystal under electron irradiation at 600°C. The initial configuration is shown in the left panel, in which the cylindrical hollow cavity with constant diameter is partly filled with a Fe_3C crystalline nanowire. After irradiation, the upper half of the tube shrank radially, applying high pressure on the encapsulated Fe_3C crystal. As a result, the diameter of the encapsulated Fe_3C crystal was reduced and the solid was squeezed through the hollow core downward along the tube axis (see the middle and right panels in Fig. 10.9). Such the downward motion of the Fe_3C crystal was due to the high pressure application. This means that irradiated MWNTs can serve as self-contracting high-pressure cells that control the geometry of rigid materials encapsulated in the hollow cavity.[a]

[a]The irradiation-driven high pressure effect was originally discovered in the study of carbon onion, i.e., assemblies of concentric spherical carbon shells [307]. It was found in 1996 that under electron-beam irradiation, carbon onions shrink toward smaller onion structures, as a result of which extreme pressure arises in the core of spherical carbon shells.

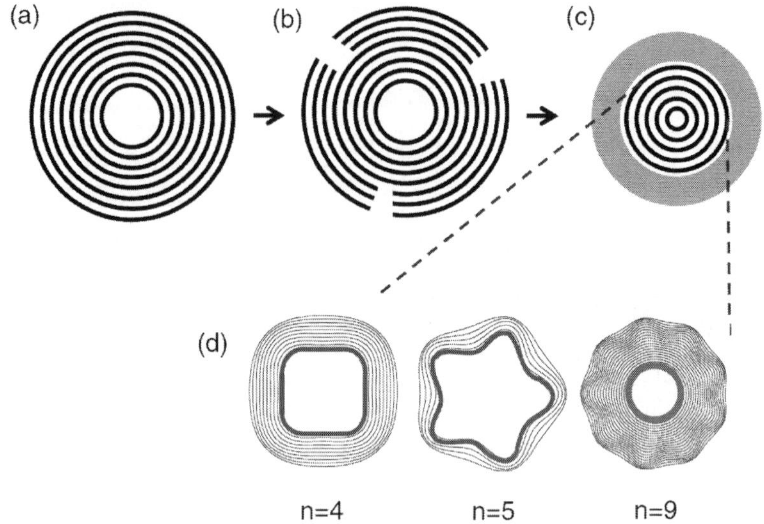

Figure 10.10 Illustration of radial shrinking hypothesis of MWNTs. (a) An $(M + N)$-walled nanotube is subjected to electron-beam irradiation. (b) Irradiation with fine resolution kicks off a portion of carbon atoms located within M ($\gg N$) outermost walls. (c) The eroded M walls are healed and radially contracting, as a result of which extreme pressure is exerted on the N innermost walls. (d) Typical corrugation modes of the pressed N-walled nanotubes are shown with the mode index n. Reprinted with permission from Shima, H., Sato, M., Iiboshi, K., Ghosh, S., and Arroyo, M., *Phys. Rev. B*, **82**(8), 085401, 2010. Copyright (2010) by the American Physical Society.

The experimental finding mentioned above leads to a conjecture that if only outmost carbon walls of an MWNT are eroded selectively by irradiation (which will be possible by using finely focused beam), then self-contraction of the outer walls exert high pressure on the encapsulated inner walls. Such high pressure application implies a novel class of cross-sectional transformation of the inner walls, as theoretically demonstrated in the recent work [335].

Figure 10.10 illustrates the situation considered. Suppose an $(M + N)$-walled nanotube under irradiation that kicks off a portion of carbon atoms located within M ($\gg N$) outer walls surrounding N inner walls. Then, self-contraction of the M walls results in high pressure that exerts on the N walls that they encapsulate. (see Figs. 10.10(b–c)). As a consequence, a wide variety of radial-corrugation patterns (i.e., circumferentially wrinkling structures) in the cross section of the N-walled nanotube. Only a few examples of the corrugation patterns are shown in Fig. 10.10(d) together with the associated mode index n representing the wave number in the circumferential direction. In Fig. 10.10(d), the two left-hand-side panels show "inward-deformation" modes with radial-corrugation mode indexes $n = 4$ and 5, and the right-hand-side panel shows an "outward-deformation" mode with $n = 9$. In the inward-deformation mode, the innermost walls exhibit significant corrugation amplitudes as compared to the outside walls. Conversely, in the outward-deformation mode, the innermost wall maintains its circular shape. Which class of modes is observed depends on the values of the innermost tube radius R and N under consideration. It was numerically confirmed that larger R and smaller N favor the inward mode with larger n [335].

10.8 Exhibition of Various Tailoring Techniques

10.8.1 *Fusion of Adjacent SWNTs*

We learned in the previous sections that when the irradiated tube is isolated, it will shrink by mending radiation-induced vacancies through atomic rearrangements, leading to surface reconstruction and dimensional changes [252]. On the other hand, if the nanotubes to be irradiated are initially assembled into bundles, they will establish links in order to satisfy dangling bonds and to minimize the strain energy by creating larger-diameter nanotubes [59].

The above argument implies that electron-beam irradiation causes nanotube fusion. For instance, it makes two adjacent SWNTs to a single larger-diameter SWNT, or a bundle of SWNTs to a single MWNT [310]. Such fusion process has been visualized in Ref. [336] (though the fusion presented there was caused by thermal activation, instead of the irradiation-induced activation.) Figure 10.11 illustrates a coalescence process of two neighboring DWNTs [336]. Two outer tubes of the DWNTs merge through a zipping mechanism, as a result of which they are completely combined into a single large outer tube with an oval shape containing two

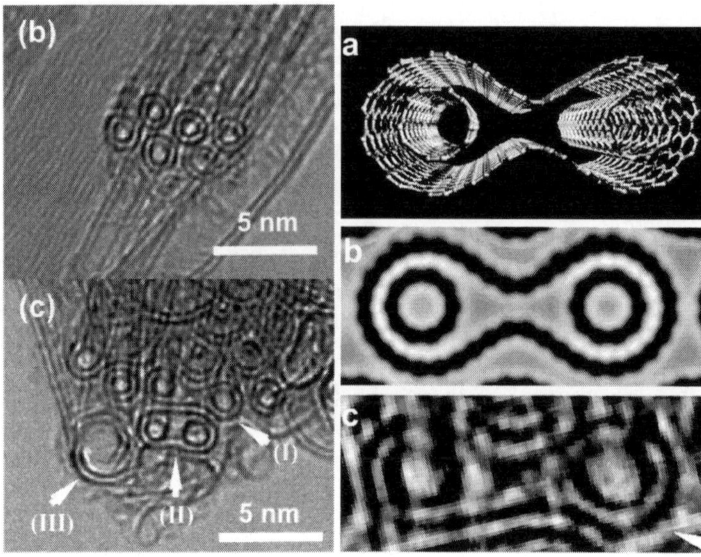

Figure 10.11 Left: (b) A typical high-resolution transmission electron microscopy (HR-TEM) image of DWNTs in a bundle state. (c) HR-TEM image of DWNTs at 2100°C. This image exhibits a sequential reconstruction process of a DWNT: (I) two outer tubes start to merge, through a zipping mechanism, (II) two outer tubes are completely combined into a single large outer tube with an oval shape containing two SWNTs, and (III) two inner SWNTs in a confined space might decompose along the inner wall of an outer shell, to form one inner single shell. Right: (a) Molecular model of two coalesced DWNTs consisting of an outer dumbbell-like tube containing two parallel individual SWNTs, (b) HR-TEM image simulation result, and (c) HR-TEM image of two SWNTs contained inside a dumbbell carbon cylinder. Reprinted with permission from Endo, M., Hayashi, T., Muramatsu, H., Kim, Y. A., Terrones, H., Terrones, M., and Dresselhaus, M. S., Coalescence of double-walled carbon nanotubes: formation of novel carbon bicables, *Nano Lett.*, **4**, 1451–1454, (2004). Copyright (2004) American Chemical Society.

SWNTs. The computer simulations [337] confirmed that coalescence is catalyzed by the presence of dangling bonds and follows a zipper-like process as presented in Fig. 10.12.

Another interesting nanotube fusion under irradiation is the transformation of an SWNT bundle into a single MWNT. Figure 10.13 shows an irradiated SWNT bundle with an electron beam of 20 nm in diameter [338]. In the irradiated area, the bundle (still

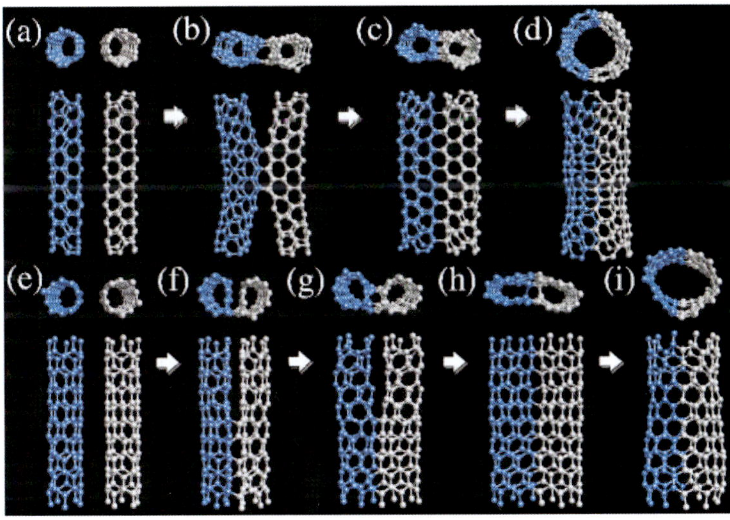

Figure 10.12 Coalescence process of the same two achiral nanotubes by colliding with each other (top and side views). The figures (a–d) are the snapshots of the reaction for (3,3)+(3,3), and the figures (e–i) are those of the reaction process for (5,0)+(5,0). The resulting structures were identified as armchair (6,0) and zigzag (10, 0) nanotube, respectively. The blue (gray) spheres represent the carbon atoms originally belonging to the left(right)-hand-side nanotube. Note that there is no defect before and after the coalescence. Reprinted with permission from Kawai T., Miyamoto Y., Sugino O., and Koga Y., *Phys. Rev. Lett.*, **89**(8), 085901, 2002. Copyright (2002) by the American Physical Society.

visible at the diagonal edges of the image) collapses and transforms into an MWNT (seen in the center of Fig. 10.13). Intriguingly, a coherent junction between SWNTs in the bundle and an MWNT can be created, which is hardly feasible in ordinary fabrication methods of the top-down style.

10.8.2 *Cutting and Bending*

Different from a spatially fixed one, a mobile focusing spot of electron beam allows us to cut a bundle of SWNTs (and of course an isolated SWNT, too), either partially or completely [333, 339]. This is shown for bundles in Fig. 10.14. The cutting speed was determined

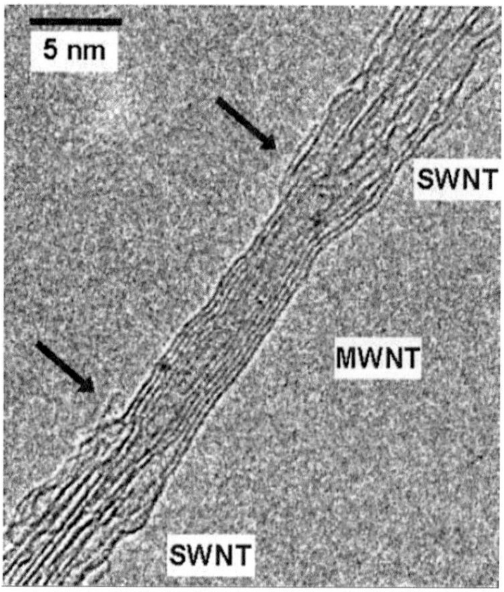

Figure 10.13 Coherent transition from a bundle of SWNTs to an MWNT. The irradiation of the bundle for 28 min leads to the transformation into an MWNT in the irradiated area indicated between the two arrows. Reprinted with permission from Li, J. X., and Banhart, F. The engineering of hot carbon nanotubes with an electron beam, *Nano Lett.*, **4**, 1143–1146, 2004. Copyright (2004) American Chemical Society.

by the beam intensity, and surprisingly, by the existence of close cuts at the ends of the tube. We see from the image that after cutting, the open ends of the tube close immediately with hemispherical caps. Hence, the cutting speed decreased when the tube was closed by a cap, since the interstitials could not escape from the irradiated area and thus the defect annihilation rate was higher [333].

Besides the cutting, the irradiation makes it possible to bend MWNTs [338] due to the removal of carbon atoms from one side of the tube (Fig. 10.14). The bending angle could be controlled precisely by the irradiation dose.

10.8.3 *Welding into Branched Nanotubes*

Another strategy for nanotube's morphology manipulation is to connect different nanotubes together by welding the joint area at

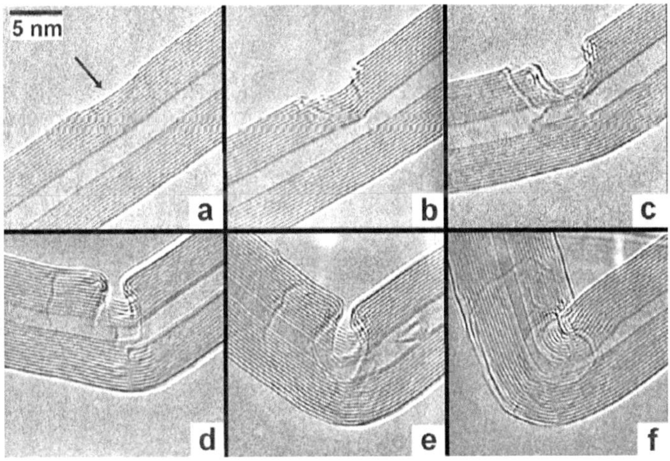

Figure 10.14 MWNTs under irradiation. The sequence shows the bending of the nanotube when the beam spot size equals the wall thickness (5 nm). Irradiation time: (a) 3 min, (b) 5 min, (c) 7 min, (d) 11 min, (e) 13 min, (f) 15 min. Reprinted with permission from Li, J. X., and Banhart, F. The engineering of hot carbon nanotubes with an electron beam, *Nano Lett.*, **4**, 1143–1146, 2004. Copyright (2004) American Chemical Society.

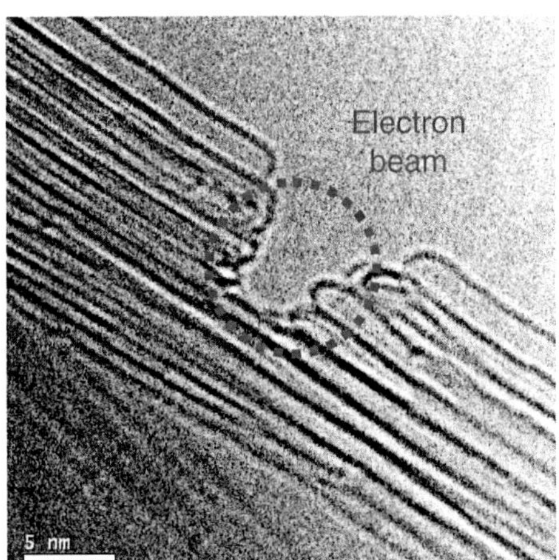

Figure 10.15 TEM picture of a nanotube bundle partially cut by the electron beam. It is evident that tubes develop caps at the cuts. Reprinted with permission from Gan, Y., Kotakoski, J., Krasheninnikov, A. V., Nordlund, K., and Banhart, F., The diffusion of carbon atoms inside carbon nanotubes, *New J. Phys.*, **10**, 023022, 2008.

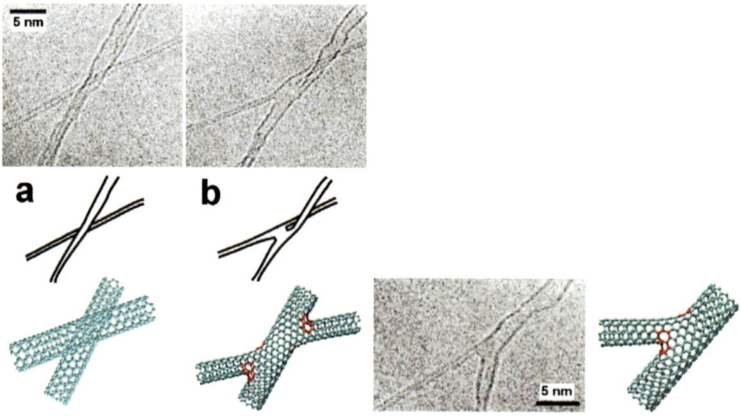

Figure 10.16 Left: (a) A thick SWNT (running from bottom-left to top-right) crossing with an individual thin SWNT. (b) The two SWNTs after electron beam irradiation, forming an "X"-shaped junction via molecular connection between them. Molecular models of each image are provided; heptagonal rings are indicated in red. Right: HR-TEM image and molecular model of a "Y"-shaped junction created following electron irradiation of an "X" structure. One of the arms of the X junction vanished due to continuous sputtering under the electron beam, and a three-terminal junction remained. Reprinted with permission from Terrones, M., Banhart, F., Grobert, N., Charlier, J.-C., Terrones, H., and Ajayan, P. M., *Phys. Rev. Lett.*, **89**(7), 075505, 2002. Copyright (2002) by the American Physical Society.

high temperature by electron beam irradiation [310, 312].[a] The left panels of Fig. 10.16 shows that two crossing SWNTs irradiated at high temperature (~800°C) are welded at the point of contact, which resulted in the formation of an X-shaped junction [342]. By using further careful electron irradiation, one arm of the X-shaped junction was able to be removed to create a Y-shaped junction [312] (see Fig. 10.16(b)). On the issue of the welding, many theoretical work have been reported [343–350] to understand its microscopic mechanism; see the literature for details.

[a]Ion beam irradiation [341] and current annealing [340] were also reported to have the ability of nanotube welding. In the latter experiment, two nanotubes were mechanically contacted and then a bias voltage was applied across the contact. If the current reached a threshold value (~6 μA), the Joule heating makes the two nanotubes join together into one carbon nanotube in a seamless manner.

Chapter 11

Internal Sliding

The higher a man's position, the more slippery his foot becomes.
— Cornelius Tacitus, in "Annales"

11.1 Potential Nano-"Trombone"

An ideal MWNT has no chemical bond between the atomistically smooth carbon layers within the cylindrically nested structure; the only inter-layer coupling comes from the weak van der Waals (vdW) interaction among constituent atoms. The loose coupling between concentric carbon cylinders alludes to a striking telescoping property whereby individual cylinders can slide and rotate with respect to each other [351]; see Fig. 11.1 for the internal sliding behavior of a double-wall carbon nanotube. The earlier experiments confirmed that both the translational [28, 351, 352] and rotational shear strengths [353] within an isolated MWNT can be less than 1 MPa, extremely small compared with other solids. This extremely low friction against the interwall motion within of MWNTs led to the idea of constructing a new set of mechanical (and electromechanical) nanodevices that are endowed with the high-speed, low-friction, and low-power-consumption

Elastic and Plastic Deformation of Carbon Nanotubes
Hiroyuki Shima and Motohiro Sato

ISBN 978-981-4310-75-8 (Hardcover), 978-981-4364-15-7 (eBook)
www.panstanford.com

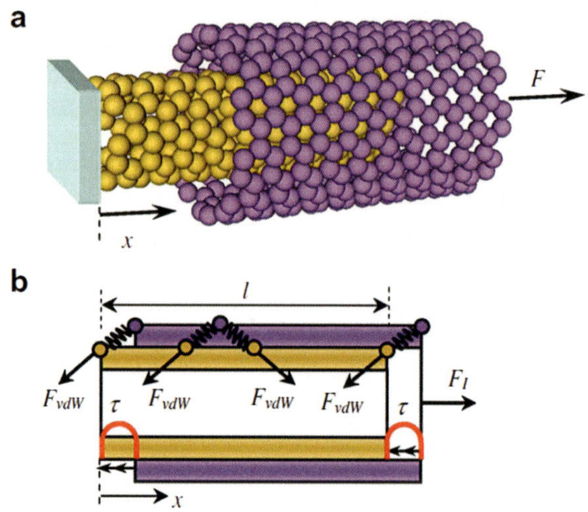

Figure 11.1 (a) Pullout of a double-wall carbon nanotube. The yellow balls represent for atoms of the inner wall while the purple ones denote those of outer wall. (b) Diagram of the vdW force F_{vdW}, the pullout force F_{l}, and stress components τ on sliding interface. Reprinted from *Carbon*, **48**(10), Li, Y., Hu, N., Yamamoto, G., Wang, Z., Hashida, T., Asanuma, H., Dong, C., Okabe, T., Arai, M., and Fukunaga, H., Molecular mechanics simulation of the sliding behavior between nested walls in a multi-walled carbon nanotube, 2934–2940, Copyright (2010), with permission from Elsevier.

characteristics: Nanoelectromechanical switches [354, 355] and nanoscale actuators[a] [356–359] are the successful examples of experimental realizations.[b]

Currently, nonetheless, there remain a lot to sort out. Readers may notice, through literature searches, that the studies on the interwall motion within MWNTs have been primarily confined to theoretical work even in the last few years [360–368]; namely, there has been to date only a few experimental demonstrations. Moreover, while prototypes have proven the utilities of the proposed interwall-motion-based devices, it is still a significant challenge how to alter the as-synthesized closed-cap structure of MWNTs

[a]An actuator is a mechanical device that is operated by the conversion of supplied energy (usually in the form of an electric current) into the mechanical motion of a system.

[b]An extensive review on the issue can be found in Ref. [521].

into a desired form suited to applications: open-edge structure (see Section 11.2) and partially peeled one (Section 11.4) are the case in point. The latter problem is particularly relevant in view of the manufacturability and commercialization of the potential devices.

With the still developing situation in mind, we are devoted in this chapter to the following three fundamental topics: lateral internal sliding, internal rotation, and associated internal oscillations of MWNTs. Emphasis will be placed mainly on the experimental evidence of those internal motions obtained thus far, together with concise theoretical arguments at need.

11.2 Low-Friction Sliding

11.2.1 *Initial Theoretical Prediction*

It all starts with the density-functional-theory-based analysis in 1993 predicting that geometrically perfect MWNTs would be the "smoothest bearings" [369]. Suppose that you pull out a few inner layers of an MWNT fixed at one end and move it back and forth. Then, you will observe a sword-in-sheath behavior in which the outer layers of the MWNT form a sleeve and the inner core forms a shaft. The remarkable consequence of the analysis [369] is that such telescoping MWNTs exhibit vanishingly small interlayer friction due to the suppression of collective stick-slip motion, similarly to superlubricity in graphite [370].

Subsequent theoretical efforts suggested that frictional properties of the telescoping MWNTs are dependent on size, temperature, velocity, and other diverse factors. However, unfortunately, no general theory addressing the internal friction of MWNTs has been succeeded thus far; this is reflected in the difficulty in treating the friction at the nanoscale. For instance, some theoretical studies insisted that a registry-dependent potential [371–373] that originates from the chirality difference between individual interacting walls should be responsible for the telescope motion. Other work considered the telescoping MWNT under radial pressure [374], concluding that the registry effect gives only a minor contribution. Instead, interaction between the "ends" of mobile core and the casing [374–376] or thermally induced fluctuations [377] were suggested to take a primary role in determination of the internal friction. Against the controversial background, continued efforts would be desired to establish the unified scenario that enables a quantitative prediction for the frictional properties of MWNTs.

11.2.2 *Experimental Observation*

After the initial prediction done in 1993, more than a half decade was required for direct experimental observation on the smooth sliding within an isolated MWNT. The first attempt to highlight the easy interwall sliding [351] was carried out in 2000 according to the process illustrated in Fig. 11.2. The setup consisted of first opening outer caps at one end of an MWNT, then exposing the inner shells (see Fig. 11.2(a) and (b)). A nanomanipulator was then brought into contact with the tip of the exposed inner tubes, and welded to them by means of a short pulse of current (Fig. 11.2(c)). This makes it possible to pull out the core of inner tubes from the external shells in a sword-in-sheath fashion, or pull in and out repeatedly as illustrated Fig. 11.2(d). Indeed in the seminal experiment [351], appropriately 20 cycles turned out to give no apparent wear or fatigue, implying the low-friction nature of the interwall motion within the MWNTs.

A similar low-friction sliding was observed in 2000 [29], too, by applying the axial tensile load on the pristine (not edge-opened) MWNTs. The observation obtained is depicted in Fig. 11.3; a sufficiently large tensile load caused a break of the outmost wall and its sliding against the other inner walls. The measured sliding friction force of nested carbon nanotubes was about 7.2×10^{-5} nN/atom, which agrees fairly with the estimate (1.5×10^{-5} nN/atom) by open-edge MWNT tests mentioned above [351].[a,b]

[a]It should be remarked that the open-edge system is dominated by only the vdW interaction, while the outmost-fractured system inevitably involves mechanical interlocking or defects at the interface caused by the initial break.

[b]Even in perfectly defectless nanotubes, thermally induced fluctuations [377] or phonon excitation due to interaction of the core with the edge of the outer casing [375] could lead to wall-to-wall friction that would scale with the length of the moving section.

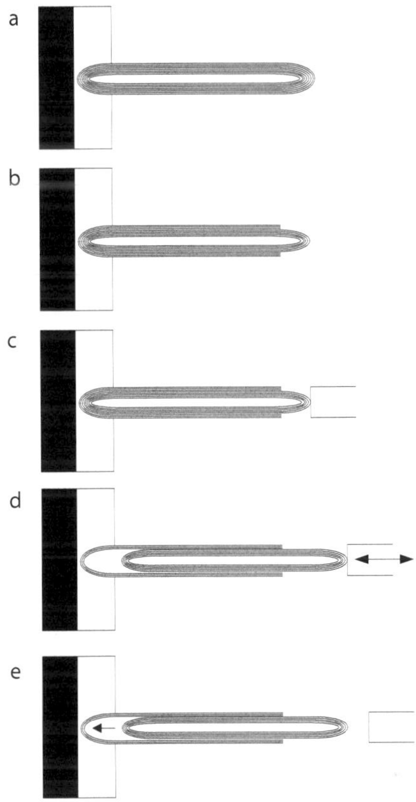

Figure 11.2 Schematic representation of the interwall sliding experiment. (a) Pristine MWNT whose one end is anchored. (b) Exposing the inner walls by opening the other end. (c) Attaching the nanomanipulator to the inner walls. (d, e) Two different classes of subsequent experiments performed. In (d), the nanotube is repeatedly telescoped while observations for wear are performed. In (e), the inner walls are released and pulled into the outer-wall housing by the attractive vdW force. Adapted from Ref. [351].

Afterward, a more sophisticated experiment in 2006 [28] revealed the interlayer sliding friction of MWNTs below the measurement limit of 1.4×10^{-6} nN/atom. This is an order of magnitude lower than the previous two estimates and also lower than the observed friction for one of the least dissipative nanoscale interfaces—the incommensurate contact between C_{60} islands and NaCl [378].

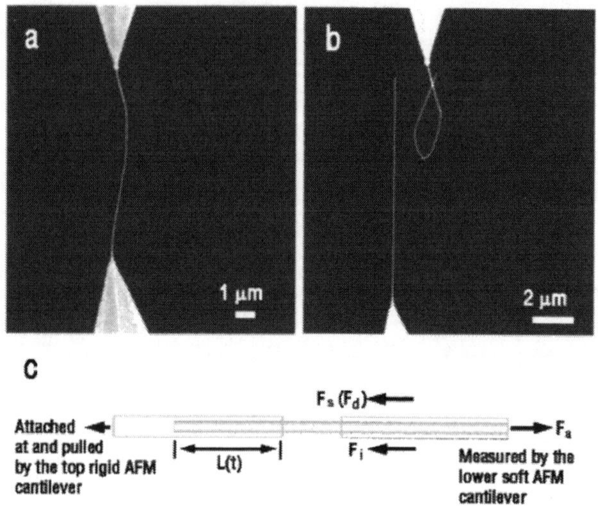

Figure 11.3 Sword-in-sheath breaking mechanism of MWNTs. (a) An MWNT attached between microscope tips under no tensile load. (b) The same MWNT after being tensile loaded to break. Notice that the overall length of the MWNT fragments left after break is clearly longer than the initial length, with in mind the curling of the top MWNT fragment. (c) A schematic showing the sliding and pulling out of internal shells from the outer shell of the MWNT. Reprinted with permission from Yu, M. -F., Yakobson, B. I., and Ruoff, R. S., Controlled sliding and pullout of nested shells in individual multiwalled carbon nanotubes, *J. Phys. Chem. B*, **104**, 8764–8767, 2000. Copyright (2000) American Chemical Society.

11.2.3 *Gigahertz Oscillator Inspired*

Another remarkable consequence from the pullout experiment [351] is the possibility of high-frequency oscillation of the extruded inner walls, implying the utility of the telescoping MWNTs as a nanoscale resonator. To be noticed is the presence of the restoring force of the vdW interaction that acts on the extruded nanotube. Suppose that the inner walls are pulled out by a manipulator, and then released at the point of largest telescopic extension. Due to the vdW-mediated attraction, the inner walls will retract into the outer casing, due to the pull of the vdW force. A simple analysis gave the estimation that the inner walls are telescoped back within on the order of 10^{-9} seconds under the experimental condition of Ref. [351]. Such a fast reaction offers gigahertz oscillator fabrication based on open-edge MWNTs [371, 379, 380], which will be discussed in Section 11.5 in more detail.

11.3 Shell-by-Shell Extraction

The easy-to-slide property of MWNTs enables to extend them in a stepwise fashion [381], resulting in the extraction of the innermost shell (i.e., a thin SWNT) without peeling the outer shells. The essence of such stepwise extraction method can be understood from Fig. 11.4. Figure 11.4(a) shows the result of moving an microscope tip along the substrate perpendicular to the axis of a long MWNT. With increasing the length of MWNT segment displaced by the tip, the total force required to overcome the adhesion of moving the MWNT with the substrate increases. When the force applied by the tip exceeds the fracture strength of the outer shells (~10–100 GPa), the inner shells cast off the skin (i.e., the ruptured outer shells) and extruded as presented in the inset of Fig. 11.4(a).

As the tip motion further continues, the extruded inner shells are subject to the next stage of extraction; the outer portion of the extruded shells is cracked when the applied load surpasses the adhesion force between the shells and the substrate. By repeating the procedure, the diameter of the extruded shells gets reduced stepwise, coming down to the innermost core shell of the original MWNT.[a] So finally, an isolated ultrathin SWNT can be provided by using the present method. The isolated SWNTs turned to have diameters as small as 0.4 nm, which can hardly be obtained in conventional synthesis techniques.

Figure 11.4(b) shows the height profile along the resulting structure. The nanotube diameter drops in a stepwise fashion, corresponding to the successive extraction of the inner shells. The

[a]The length of the extracted nanotube can be controlled by the scanning tip manipulation as follows: When the extracted nanotube comes into contact with the substrate, it will rupture in a fashion analogous to that just described for the initial extraction step. On the other hand, a much greater extension of the inner shells is possible if the tip guides the nanotube above the surface so as to get contactless, thus eliminating adhesion with the substrate.

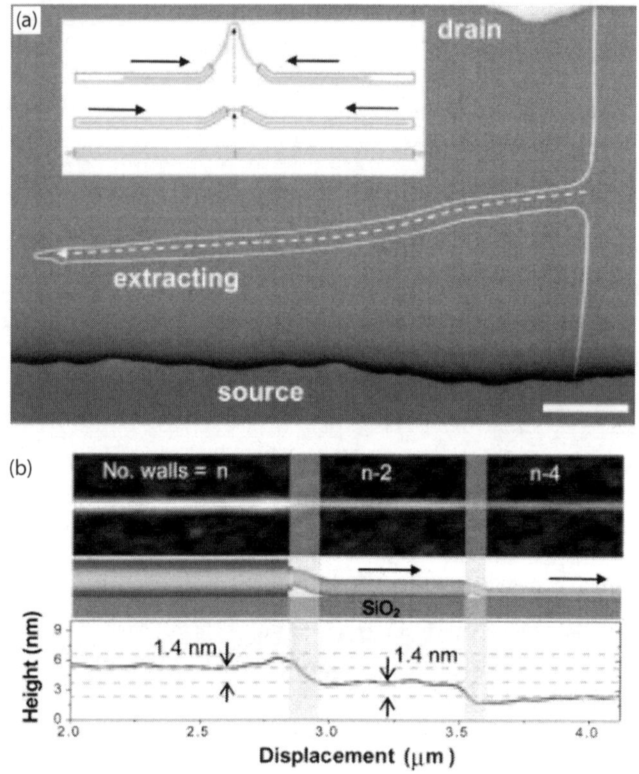

Figure 11.4 Microscope images and a height profile showing the shell-by-shell extraction process of MWNTs. (a) A manipulated MWNT formed by moving the microscope tip vertical to the nanotube axis. Inset illustrates how the tip breaks the outermost shells and causes the extrusion of the inner shells of the MWNT. Scale bar indicates 10 μm. (b) A height profile showing the stepwise decrease of the diameter along the extracted nanotube. Each successive structure corresponds to a decrease of the hollowed-out nanotube by two shells. Reprinted with permission from Hong, B. H., Small, J. P., Purewal, M. S., Mullokandov, A., Sfeir, M. Y., Wang, F., Lee, J. Y., Heinz, T. F., Brus, L. E., Kim, P., and Kim, K. S. (2005). Extracting subnanometer single shells from ultralong multiwalled carbon nanotubes, *PNAS*, **102**, 14155–14158. Copyright (2005) National Academy of Sciences, U.S.A.

decreasing step typically equals to ~1.4 nm (see Fig. 11.4(b)), which corresponds to approximately four times the intershell spacing (~0.34 nm) in MWNTs, indicating that the extraction process leaves behind hollowed-out nanotubes consisting of pairs of shells. The preference for producing double-shell structures after extrusion was attributed in Ref. [381] to an enhancement of the friction between the two outermost nanotubes associated with the cross-sectional deformation of the outer shell due to the vdW coupling with the substrate [382], though no quantitative interpretation was obtained.

11.4 Internal Rotation

In MWNTs, the ultra-low-friction between concentric walls makes it possible to rotate them almost freely with each other around the tube axis. Such expected frictionless rotation has inspired the fabrication of "nanomotors" with much smaller dimensions than the conventional mechanical motors, opening the development a new class of mechanical (and often electro- or opto-mechanical) devices.

The pioneering work on this issue, published in 2003 [356], demonstrated a rotational nanoactuator. Figure 11.5 designs the concept of the proposed actuator. A rectangular metal plate (labeled by "R"), serving as a rotor plate, is attached to a suspended support shaft (i.e., the MWNT). The MWNTs' ends are embedded in electrically conducting anchors ("A1" and "A2"); hence, the MWNT functions as both an electrical wire and a rotational bearing. The rotor plate is surrounded by three fixed electrodes. Four independent voltage signals, one to the rotor plate and three to the electrodes, can be applied to control the rotation of the plate.

Note that this assembly remains to be a torsional spring [383–385] if nothing is done. In order to obtain a low-frictional bearing, very large torque was applied successively to the plate until the outer shells fractured mechanically. Once the outer shells break partially or completely, a dramatic increase in the rotational freedom of the rotor plate can be attained, since a short section of the outer shell between the plate and the anchors was removed. In the "free" state, the rotor plate was still held in position axially by the intact nanotube core shells, but its rotational angle could be controlled

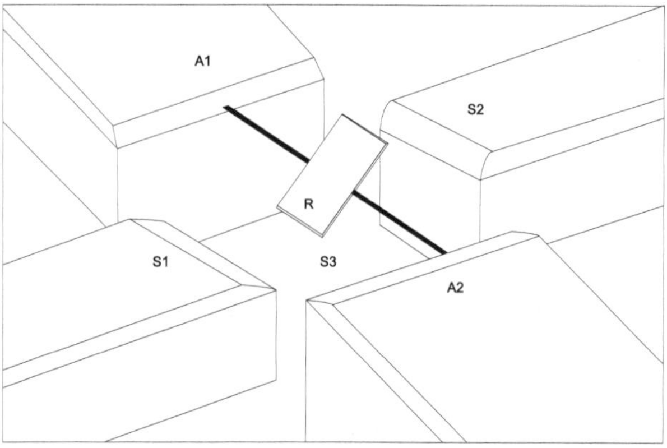

Figure 11.5 Conceptual drawing of nanoactuator. A metal plate rotor (R) is attached to an MWNT that acts as a support shaft and is the source of rotational freedom. Electrical contact to the rotor plate is made via the MWNT and its anchor pads (A1, A2). Three electrodes, two on the SiO_2 surface (S1, S2) and one buried beneath the surface (S3), provide additional voltage control elements. Adapted from Ref. [356].

using an appropriate combination of signals from electrodes, to any arbitrary angle between 0° and 360°. The free rotation of the actuator was realized by varying the applied voltages appropriately. By reversing the sequence of voltages, the rotor plate rotation could be reproducibly reversed.[a]

It is important to note that in the above-mentioned fabrication, the rotational freedom is obtained by mechanically fatiguing, wherein the structural modifications brought to the MWNT are not controllable. This drawback was overcome in 2004 by attaching the rotating plate to an inner shell, not the outmost one, of a pre-formed MWNT [353]. Figure 11.6(a) shows the setup in which the assembly of the plate and the attached inner shell turn around the tube axis while outer shells are fixed to two anchor pads. The plate cannot slide in the direction of the tube axis because it is held

[a] It was claimed in Ref. [356] that the proposed actuator is the first true MWNT-based nanoelectromechanical device, in the sense that it fully integrates electronic control and mechanical response; this distinguishes it from previous related devices that require relatively large and complex external control systems to achieve operation [11, 351, 352, 383, 384].

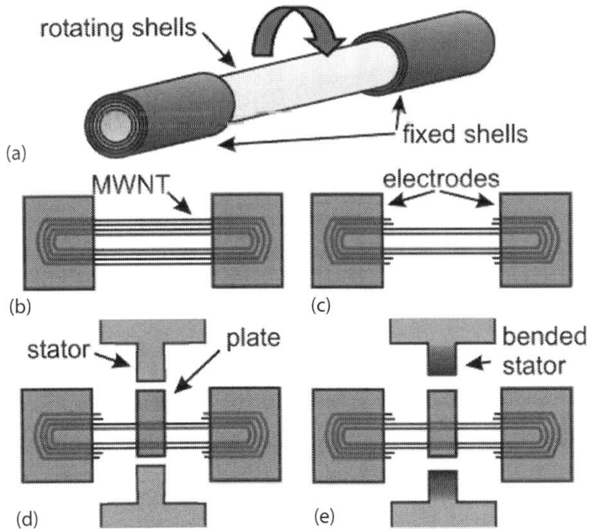

Figure 11.6 (a) Schematic of rotating inner shells inside the fixed outer shells. (b) An MWNT contacted to two conducting anchor pads. (c) Removing several shells between the contacts to gain access to a selected inner shell. (d, e) A plate and two stator electrodes fabricated. Reprinted with permission from Bourlon, B., Glattli, D. C., Miko, C., Forro, L., and Bachtold, A., Carbon nanotube based bearing for rotational motions, *Nano Lett.*, **4**, 709–712, 2004. Copyright (2004) American Chemical Society.

by the shell caps.[a] In the measurement [353], the static friction against rotational sliding was estimated to be $\sim$0.85 MPa, which is comparable to that for transversal sliding and for high-quality crystalline graphite [388].

11.5 Telescopic Oscillation

11.5.1 *Mechanism*

When extruding inner core shells from an MWNT in the axial direction, the core feels the restoring force from the surrounding

[a]The fabrication of such a device has been made possible with the electrical breakdown technique [357, 386, 387], which engineers the MWNT by opening access to the inner shells.

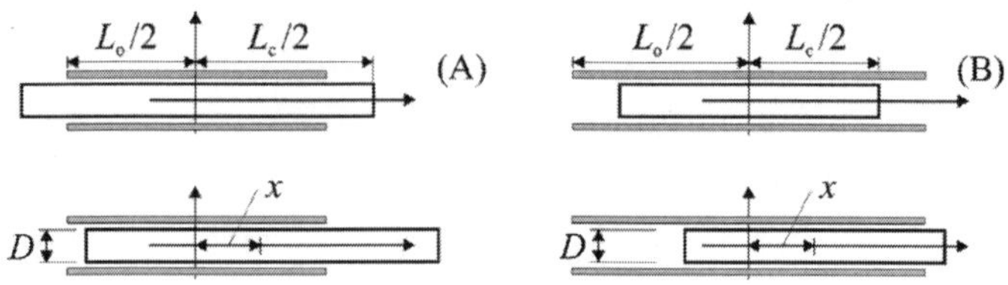

Figure 11.7 Schematic illustrations of an MWNT with a core extruded from the outer shells. (a) The core is longer than the outer shells; (b) The core is shorter than the outer shells. Reprinted with permission from Zheng, Q., Liu, J. Z., and Jiang, Q., Excess van der Waals interaction energy of a multiwalled carbon nanotube with an extruded core and the induced core oscillation, *Phys. Rev. B*, **65**, 245409, 2002. Copyright (2002) by the American Physical Society.

outer shells through the vdW coupling at the interface. Hence, the retraction energy causes the core to oscillate with respect to its fully retracted position where the vdW potential energy is minimized. This phenomenon, if experimentally achievable, leads to the creation of nanomechanical systems of operating frequency beyond one gigahertz, which is beyond the reach of existing micrometer-scale oscillation devices.

An expected oscillating behavior of the extruded MWNT is schematically depicted in Fig. 11.7, in which both ends of the outer shells are open. The core is initially positioned at the center in equilibrium as presented in the upper panels and then extruded artificially by a displacement x through the application of an external force as depicted in the lower panels. Once the extruded core is released, it is fully retracted into the outer shells where the vdW potential energy is minimized and the kinetic energy is maximized. After it passes over the position of minimum potential energy, the restoring force reverses its direction to decelerate the motion. Finally, once the core reaches the maximum extrusion from the opposite end, the motion reverses, which lead the lateral oscillation of the inner shells.

11.5.2 *Oscillation Frequency Estimation*

Estimating the frequencies of such oscillators requires calculating the restoring force resulted from the excess vdW energy due to extrusion [389]. To obtain analytic expressions, we can replace the discrete distribution of atoms on the carbon wall with a continuous distribution of the same atom density $\rho = 4\sqrt{3}/(9a^2)$ [390], where $a = 0.142$ nm is the bond length.

According to the argument in Ref. [389], denoted by x the separation distance of the center of the core from the center of the outer shell due to extrusion (see Fig. 11.7). The surface area of the portion of the core inside the outer shell is given by $S(x) = \pi D[(L_c + L_o)/2 - x]$, where D is the core diameter and L_c (L_o) is the length of the core (outer) shells. When the entire core shells are enclosed within the outer shells (i.e., $x < |L_c - L_o|/2$), the vdW energy, $U(x)$, between the n_c core shells and the n_o outer shells is independent of x as proved in Ref. [389]. Otherwise, when a portion of the core is extracted from one edge of the outer shells, we obtain

$$U(x) = \rho \times S(x) \times \Pi_{n_c,n_o}(D), \quad \left(\text{for } x > \frac{|L_o - L_c|}{2}\right) \tag{11.1}$$

$$\Pi_{n_c,n_o}(D) = \sum_{j=0}^{n_c-1} \frac{D - 2sj}{D} \left[\sum_{k=1}^{n_o} \Phi_{j+1,k}(D)\right], \tag{11.2}$$

where $s = 0.34$ nm is the wall separation. In Eq. (11.2), $\Phi_{j+1,k}(D)$ represents the average value of the vdW energy between all the atoms of the kth outer shell (counted from inside to outside) and an atom of the jth core shell (counted from the outside to inside). Since $\Phi_{j+1,k} < 0$ under the present situation, the factor $\Pi_{n_c,n_o}(D)$ in Eq. 11.2 is always negative for arbitrary D.

Due to the simple dependence of $U(x)$ on x, the vdW restoring force F_{vdW} is readily derived as

$$F_{\text{vdW}} = -\frac{dU(x)}{dx} = \begin{cases} \rho\pi D \Pi_{n_c,n_o}(D), & (\text{for } x > |L_o - L_c|/2) \\ 0. & (\text{otherwise}) \end{cases} \tag{11.3}$$

The result indicates that the core feels a linear attractive force during the extrusion, while it shows free motion when the entire core is inside the casing. Finally, applying Newton's second law expressed by $m(d^2x/dt^2) = F_{\text{vdW}}$ with m being the total mass the core, we obtain the oscillation frequency of the extruded inner core tube.

We are ready to evaluate the oscillation frequency f of the extracted core. It follows that the core mass m is proportional to the core length L_c while the restoring force is independent of L_c. This means that the shorter L_c gives the higher f as understood from Eq. (11.3). As an example, we can set $L_c = 100$ nm, $n_c = 4$, $n_o = 5$, $D = 4$ nm by referring to the existing experiment [351]. In addition, we take the initial extrusion length to a quarter of L_c. By choosing the conventional 6-12 Lennard–Jones potential[a] for the carbon atomic interaction (the choice is relevant to determine the functional form of $\Pi_{n_c,n_o}(D)$), we will obtain $f \sim 1.4$ GHz [389]. If the core consists of a single shell, f becomes even higher because of the smaller m.

Further increase in f may be obtained if one end of the outer shells is closed. In this case, the core should be subjected to a huge repelling force immediately after it passes over the position of minimum potential energy, and it would extrude again because it should have a velocity toward the open end when it quickly returns, driven by the repelling force, to the position of minimum potential energy. Therefore, the oscillation frequency would be nearly twice the frequency of the nanotube with two open ends of the outer shells.

11.5.3 *Perspective*

The argument in Section 11.5.2 is somewhat oversimplified, taking neither account of the energy dissipation nor the thermal fluctuation. Key issues toward the experimental realization of the MWNT-based oscillator are how to sustain the oscillation in a controllable way as well as how to initialize the oscillation out of the equilibrium MWNT structure. Different strategies have been proposed thus far [367, 391–393], most of which are limited in consideration to only "double"-wall nanotubes, not to general MWNTs, due to the ease of dealing with in computation. These findings may serve as a stepping stone toward both further development in theory and the first experimental proof of the gigahertz oscillation in future.

---

[a] The Lennard–Jones potential is a simple model that approximates the interaction between a pair of neutral atoms or molecules. It is expressed by $V_{LJ}(r) = 4\epsilon\left[(\sigma/r)^{12} - (\sigma/r)^{6}\right]$, where ϵ represents the potential well depth and σ is the separation between atoms at which $V_{LJ} = 0$.

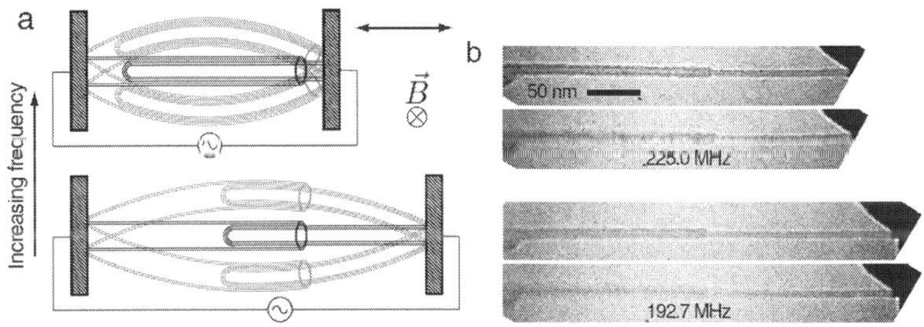

Figure 11.8 (a) Schematic drawing of our tunable resonator. A specially prepared MWNT is suspended between a stationary contact and mobile electrode. Varying the length of the nanotube through the controlled telescoping of the inner nanotube core from the outer nanotube shell tunes its resonant frequency. Operating the device in an external magnetic field, B, allows actuation with alternating current via the Lorentz force. (b) Micrographs show the device in action. The dark region on the right is the mobile contact, while the stationary clamp is more than 200 nm off the image to the left (the bulge pictured at the left of the image is actually the beginning of an MWNT bundle). The top two images show the nanotube at one extension off resonance (sharp) and on resonance (blurred) with a resonant frequency f = 225.0 MHz. The bottom two images show an extended nanotube with a lowered resonant frequency (192.7 MHz). Reprinted with permission from Jensen, K., Girit, C., Mickelson, W., and Zettl, A., Tunable nanoresonators constructed from telescoping nanotubes, *Phys. Rev. Lett.*, **96**, 215503-1–4, 2006. Copyright (2006) by the American Physical Society.

11.6 Plucking a Nano-"Guitar" String

Different from the lateral resonator discussed in Section 11.5, a distinct class of resonator can be exploited by harnessing the telescoping property of MWNTs in a new fashion. It is a tunable nanoscale resonator exhibiting transverse oscillation, not lateral one, which operates at a desired frequency in principle over a range of 100 MHz [394]. Figure 11.8(a) is a schematic drawing of the tunable nanoresonator. A pre-formed MWNT, wherein a set of thin capped core shells are inserted into thick open-edged shells, is suspended between a stationary metal electrode and a mobile contact. In the top image of Fig. 11.8, for example, the resonator is fully retracted and has a relatively high resonant frequency. In the bottom image, in contrast, the resonator is extended and consequently has a lower resonant frequency. To excite the transverse oscillation, the device is subjected to an external magnetic field ($\sim$1.3 T) and an alternating electrical current is applied through the MWNT; the resulting Lorentz force [395] in the direction vertical to both the tube axis and the magnetic field generates the mechanical vibration of the suspended MWNT.

Micrographs in Fig. 11.8(b) show the tunable nanoresonator in action. The dark region on the right is the mobile contact, while the stationary clamp is more than 200 nm off the image to the left. Note that the bulge on the nanotube shown in the left portion of the image is not the clamp but rather the beginning of an MWNT bundle. The first two images show a 558 nm-long nanotube before resonance (sharp) and during resonance at 225.0 MHz (blurred). The final two images show the nanotube after the inner nanotube has been telescoped out 50 nm. The resonant frequency has shifted downward to 192.7 MHz.

Figure 11.9 gives resonant frequency versus beam extension for four different samples having Young's modulus indicated in the plot. As expected, extended nanotubes produce lower frequencies. Also, each device covers a relatively wide range of frequencies, and together the devices span nearly the entire spectrum from 30 MHz to 300 MHz. Apparent in the graph is the extreme sensitivity of resonant frequency to telescoping extension, more than 1 MHz/nm for one device, suggesting possible application as a precision distance or position sensor or strain gauge. Relatively high quality factors[a] (up to 1000) presented in the inset indicate that sliding friction between telescoping sections of our resonator is an insignificant source of dissipation.

Euler–Bernoulli beam theory describes how frequency varies with beam extension. The nanotube beam is treated as a continuum, elastic medium subject to the differential equation:

$$\frac{\partial^2}{\partial x^2}\left(YI\frac{\partial^2 y}{\partial x^2}\right)-\frac{\partial}{\partial x}\left(T\frac{\partial y}{\partial x}\right)=-\rho A\frac{\partial^2 y}{\partial t^2} \qquad (11.4)$$

[a]The quality factor (or Q factor) is a dimensionless parameter that describes how damped an oscillator is. It is defined by $Q \equiv E_0/\Delta E$, where ΔE is the energy loss per one oscillation period and E_0 is the energy supplied to the oscillator per period. A higher (or lower) value of Q indicates under- (over-)damping, thus a lower (higher) rate of energy loss.

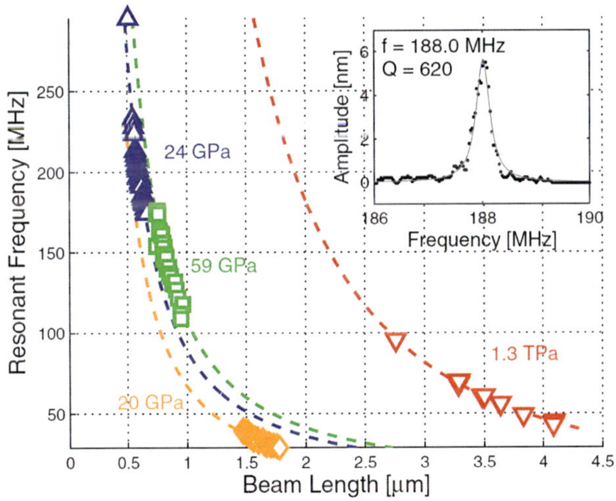

Figure 11.9 Tuning curves for four nanoresonator devices. A theoretical model provides a good fit to the data and yields reasonable values for the effective Young's modulus of each device. The inset shows a typical resonance peak with a Lorentzian fit. Reprinted with permission from Jensen, K., Girit, C., Mickelson, W., and Zettl, A., Tunable nanoresonators constructed from telescoping nanotubes, *Phys. Rev. Lett.*, **96**, 215503-1–4, 2006. Copyright (2006) by the American Physical Society.

where $y(x)$ is the transverse displacement of the nanotube along its length, Y is Young's modulus, I is the areal moment of inertia, T is the tension, ρ is the density, and A is the cross-sectional area. For a cylindrical beam with outer and inner radii, r_{out} and r_{inn}, $I = \pi(r_{\text{out}}^4 - r_{\text{inn}}^4)/4$. Applying the boundary conditions of a doubly clamped system with beam length l, $y(0) = y(l) = 0$ and $y'(0) = y'(l) = 0$, and solving the equation for the resonant frequency of the fundamental mode gives

$$f_0 = \frac{22.4}{2\pi l^2}\sqrt{\frac{YI + 0.024Tl^2}{\rho A}}. \qquad (11.5)$$

Using Eq. (11.5) to calculate f_0, the experimental tuning data in Fig. 11.9 could be fitted. In fact, the data follow the curves well and give reasonable values for Young's modulus of an MWNT (1.3 TPa) [11] or MWNT bundles (20 GPa, 24 GPa, and 59 GPa) [396].

Chapter 12

Unzipping

You can't make an omelette without breaking eggs.
— Proverb

12.1 Toward Graphene Nanoribbon Production

A carbon nanotube can be viewed as a graphene sheet rolled up into a seamless cylinder. This implies, conversely, that longitudinal cutting (i.e., unzipping) of nanotubes leads to a set of one-atom-thick graphene sheets having very narrow widths and straight edges. Such a thin strip of carbon sheet, termed a graphene nanoribbon, was theoretically predicted to conduct electrons well,[a] being transparent and mechanically strong. Hence, it has been touted as a promising material for use in electronic devices such as field-effect transistors, which form the basis of solar cells, displays, and microchips in computers. To understand and fully make use of its interesting properties, methods for large-scale production are indispensible. Until recently, however, graphene nanoribbons have proved harder

[a]A graphene sheet is a zero-gap semiconductor if it is wide enough: an electronic state precisely between a semiconductor and a metal. However, when it is made into thin strips with straight edges (i.e., nanoribbons), it can have a bandgap and shows semiconducting properties [397].

Elastic and Plastic Deformation of Carbon Nanotubes
Hiroyuki Shima and Motohiro Sato

ISBN 978-981-4310-75-8 (Hardcover), 978-981-4364-15-7 (eBook)
www.panstanford.com

to produce than nanotubes; early techniques based on lithographic patterning [398–402], sonochemical methods [403, 404], chemical vapor deposition (CVD) [405, 406], and nanocutting [407–409] could not make nanoribbons in large amounts or with controlled widths.

In the last years, several novel methods have been reported for creating graphene nanoribbons by unzipping nanotubes lengthwise in a controlled way [410, 412–415]. Unlike previous techniques, the unzipping methods enable to make nanoribbons uniform in width along their lengths, none being wedge shaped, owing to the quasi-one-dimensional nanotube templates with uniform diameter. Furthermore, the idea of unzipping nanotubes to form nanoribbons appears to be extremely powerful in terms of mass production. This is mainly because it is possible nowadays to generate hundreds tons of carbon nanotubes per year (or more) using the CVD synthesis. Therefore, by unzipping nanotubes, the nanoribbons could also be mass produced.

The unzipping processes suggested thus far can be classified as follows (Fig. 12.1):

(1) solution-based oxidation that breaks C=C bonds in the nanotubes [410]
(2) plasma etching of nanotubes placed on a polymer substrate [412]
(3) intercalation of NH_3 and Li that induces outer-wall exfoliation [413]
(4) catalytic approach in which metal nanoparticles act as scissors [414]
(5) high electric current-based burning [415]

Among these five methods, the first three were suggested nearly at the same time in 2009. An important difference between the three is the width of nanoribbons obtained. Methods (1) and (3) provide relatively broad nanoribbons (>10 nm), which could exhibit functionalized groups at the edges that affect their chemical and electronic properties. In addition, such methods have still little control on the structure of the edges of the nanoribbons produced. In contrast, method (2) allows the production of very narrow

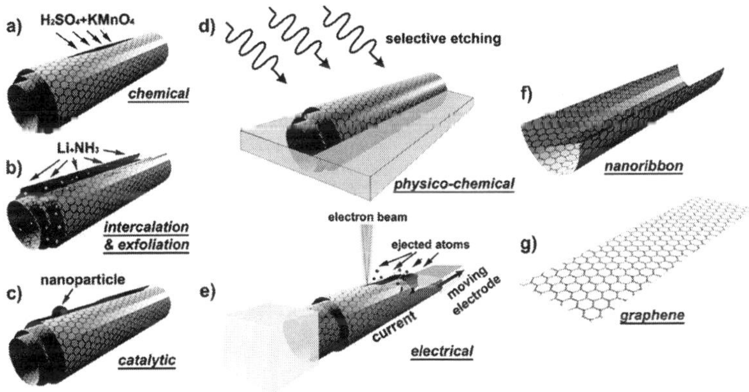

Figure 12.1 Different ways of unzipping nanotubes to yield graphene nanoribbons: (a) Acid reactions that break successively C=C double bonds in the cylindrical walls [410]; (b) Intercalation by liquid NH_3 and Li followed by exfoliation using HCl and heat treatments [413]; (c) Catalytic approach, in which metal nanoparticles cut the nanotube longitudinally like scissors [414]; (d) Ar plasma etching of the nanotubes partially embedded in a polymer matrix [412]; (e) High electric current burning [415, 421]. The resulting structures are either (f) nanoribbons partially folded or (g) those with flat shape. Reprinted with permission from Terrones, M., Sharpening the chemical scissors to unzip carbon nanotubes: crystalline graphene nanoribbons, *ACS Nano*, **4**(4), 1775–1781, 2010. Copyright (2010) American Chemical Society.

graphene (<10 nm), resulting in a sample of all semiconducting nanoribbons, but seems hardly scalable for the moment. More recently, alternative methods (4) and (5) have been reported, but the resultant quality and yield of nanoribbons are still unknown. It is thus fair to say that at this moment, a method capable of producing large amounts of high-quality nanoribbons is still to be explored, despite significant advancement proposed in the past years.

The remainder of this chapter outlines each of the five unzipping processes mentioned above. It should be borne in mind that this topic is cutting edge; therefore, other newer unzipping methods may already have been proposed when the reader gets hold of this book.

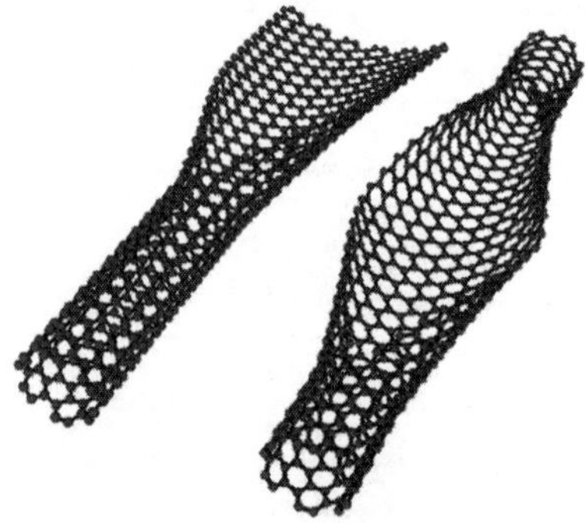

Figure 12.2 Geometry of two partially unzipped nanotubes. (Left) A (6,6) armchair nanotube unzipped into a (12,0) zigzag graphene nanoribbon, making a nanotube/graphene-ribbon single junction. (Right) The same nanotube unzipped in its central part, yielding a zigzag nanoribbon quantum dot connected to armchair nanotube contacts. Reprinted with permission from Santos, H., Chico, L., and Brey, L., *Phys. Rev. Lett.*, **103**(8), 086801, 2009. Copyright (2009) by the American Physical Society.

12.2 Acid Reaction Method

The first is a solution-based oxidative method [410] that uses a combination of $KMnO_4$ and H_2SO_4 to rip the nanotubes along a single axis (Fig. 12.2(a)) [411]. The mechanism relies on the oxidation of C=C double bonds at an undefined point along the cylindrical carbon wall (panel 2 in Fig. 12.3). The oxidation creates two neighboring C=O double bonds (so-called a diketone defect[a]) within the wall (panel 3). The resulting diketone makes adjacent C=C bonds more susceptible to oxidation. Accordingly, sequential cleavage unzips nanotubes to create highly oxidized graphene nanoribbons (panels 4 to 6). The opening of the nanotubes could occur in a linear longitudinal cut or in a spiraling manner, depending upon the initial site of attack and the chiral angle of the nanotube.

Transmission electron microscopy (TEM) analysis performed in Ref. [410] evidenced the successful synthesis of nanoribbons by oxidizing MWNTs with a diameter of 40–80 nm and 15–20 inner

[a]A diketone is a molecule containing two ketone groups; a ketone is a compound based on one C=O bond. The simplest example of diketone is diacetyl, represented by H_3C-(CO)-(CO)-CH_3, which involves two C=O bonds.

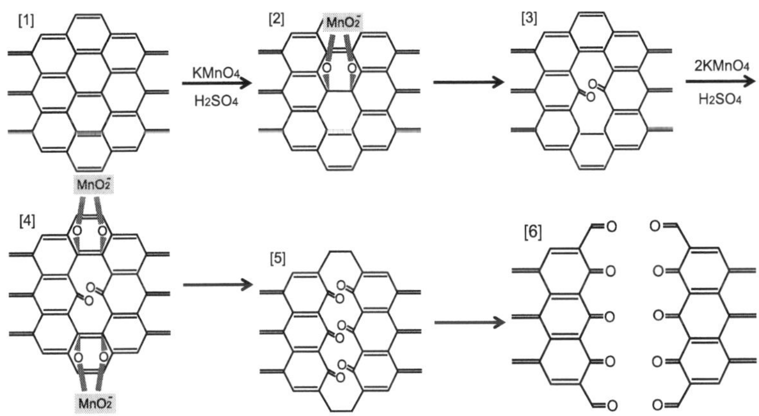

Figure 12.3 The chemical mechanism of unzipping through oxidation by $KMnO_4$ and H_2SO_4. Adapted from Ref. [410].

nanotube layers. The nanoribbons obtained are wide more than 100 nm but easier to make in large amounts. The MWNTs used in the synthesis were produced from a CVD process; although the same treatment was attempted on laser-oven-produced MWNTs, fewer nanoribbon-like structures were detected [410].

12.3 Plasma Etching

A different method to unzip MWNTs is "plasma etching." To remove a longitudinal strip of carbon atoms from the side walls of nanotubes, pristine MWNTs with a diameter of ~4–18 nm synthesized by arc discharge were deposited onto a Si substrate. Next, a polymer (PMMA) film with 300 nm-thickness was coated on top of the MWNTs. After baking, the PMMA film was peeled off; then the MWNTs turned out to be partially embedded in the polymer film. The film was then exposed to Ar plasma for various time periods. In the plasma etching process, the PMMA film acts as a mask, partially protecting the MWNT but allowing the most exposed part of the side wall to be removed with argon plasma. As a consequence, the top side walls of MWNTs were etched faster and removed by the plasma. After etching, the PMMA film was removed, leaving graphene nanoribbons.

Using this etching method, we can obtain single-, bi-, and multilayer nanoribbons as well as those with inner nanotube cores by tuning the diameter and number of layers of the starting MWNT and the etching time. For 10 seconds of Ar plasma etching, for instance, approximately 20% of the MWNTs was converted into single- or few-layer graphene nanoribbons. Other products were multilayer nanoribbons or nanoribbons with nanotube cores. Within the resolution of atomic force microscopy (AFM), the edges of the obtained nanoribbons were very smooth. Another important advantage of this method is that the widths of our nanoribbons mostly fell into the range 10–20 nm, which is narrower than that of nanoribbons made by the oxidation method described in Section 12.2.[a]

12.4 Intercalation-Induced Exfoliation

The third approach to be mentioned is based on intercalation into open-ended MWNTs with Li^+ in liquid NH_3. In this situation, intercation causes expansion of wall–wall separations in concentric tubules; beyond the threshold, the outer walls are ruptured and exfoliated graphitic ribbons and stacked sheets. The discovery of such intercalation-induced exfoliation was triggered by the study of "nanotube salts" [417]. It is well known that SWNTs dispersed in liquid NH_3 with Li form nanotube salts, in which SWNTs absorb electrons released by Li, and so becoming able to react readily with other chemical compounds [418]. While employing this functionalization method to MWNTs, it was found that partially unwrapped and exfoliated nanotubes were obtained [413]. These unexpected results motivated the study of the unzipping phenomenon in more detail, leading to the development of MWNT unzipping toward nanoribbon production.

The proposed mechanism is illustrated in Fig. 12.4. Intercalation starts by electrostatic attraction between the negatively charged MWNTs and NH_3-solvated Li^+. Structural defects on the opened

[a] It was demonstrated that highly aligned, narrow nanoribbons can be built by applying the plasma etching method to an aligned array of SWNTs [416].

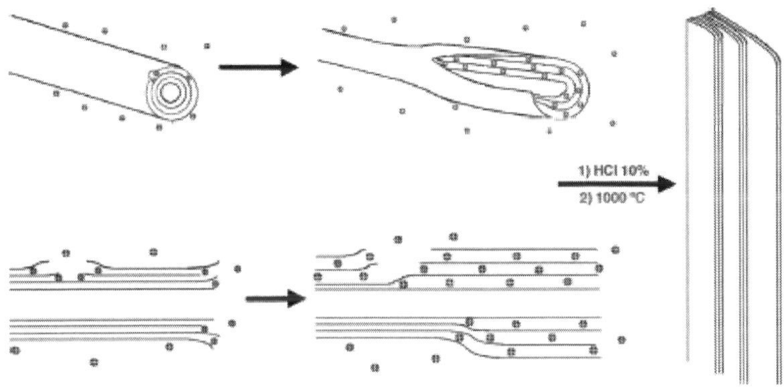

Figure 12.4 Proposed mechanism for intercalation and unwrapping of MWNT. Three-dimensional representation (above) and cross-sectional view (below) of the initial stages are shown. The dots represent NH_3-solvated Li^+, which are attracted by the negatively charged MWNTs. The intercalation expands interlayer spacing almost twice and the stress ruptures the walls. After HCl treatment, the tubes are exfoliated. After thermal treatment further exfoliation occurs together with some recrystallization of graphitic planes, producing nanoribbons. Reprinted with permission from Cano-Márquez, A. G., Rodriguez-Macías, F. J., Campos-Delgado, J., Espinosa-González, C. G., Tristán-López, F, Ramírez-Gonzalez, C. G., Cullen, D. A., Smith, D. J., Terrones, M., and Vega-Cantu, Y. I. (2009). Ex-MWNTs: Graphene sheets and ribbons produced by lithium intercalation and exfoliation of carbon nanotubes, *Nano Lett.*, **9**(4) 1527–1533. Copyright (2009) American Chemical Society.

tips act as starting points for the unwrapping, and also facilitate insertion of Li^+ ions into interwall spacings. The intercalation expands interlayer spacing almost twice and the stress ruptures the walls. The exfoliation takes place when the material was treated with HCl solution, used to remove residual Li. During the acid treatment, HCl reacts with Li ions and neutralizes NH_3, which further opens the nanotubes. Still some species remain intercalated, possibly a combination of Li-NH_3 complexes and Li without NH_3. By applying thermal treatment, therefore, their sudden removal with the aid of thermal excitation results in further exfoliation. This opens completely many nanotubes, producing graphene nanoribbons.

12.5 Catalytic Cutting

Shortly before the boost of nanotube unzipping in 2009, a controlled cutting of graphene sheets using metallic nanoparticles was demonstrated in 2008 by two independent groups [407, 419]. In the processes, nanoparticles work as a knife that cuts the sheet with atomic precision; the cutting proceeds via catalytic hydrogenation of the carbon atoms in an Ar-H atmosphere, represented by $C+2H_2 \rightarrow CH_4$, which chucks out a sequence of C atoms from the sheet. These facts imply the capability of longitudinal cutting of nanotubes by using metal nanoparticles as catalytic entities.

Such catalytic cutting of MWNTs was actually demonstrated in 2010. In the synthesis, pristine MWNTs and nitrogen-doped ones (CN_xMWNTs) were unzipped using either Co or Ni nanoparticles so as to form graphene nanoribbons. The process consists of catalytic hydrogenation of carbon, in which the metal particles cut sp^2-hybridized carbon atoms along nanotubes, which results in the liberation of hydrocarbon species. Observations have revealed the presence of unzipped nanotubes that were cut by the nanoparticles, as well as the presence of partially open carbon nanotubes. The cutting direction and depth are mostly determined by the particle size and the number of step edges where the nanoparticle nucleates.

Figure 12.5 shows SEM micrographs of the cutting process. Panels (a), (b), and (d) in the figure exhibit metal nanoparticles nucleated on the surface of CN_xMWNTs (see encircled areas). Chemically driven small nanoparticles traveled on the surface of the nanotubes creating mostly random cuts on the outer graphitic layers, via catalytic hydrogenation of the graphene sheets (panel (a)). Larger metal nanoparticles ($\sim$40 nm) traveled along the axis of some carbon nanotubes, performing deeper cuts (panels (b-d)) that opened longitudinally MWNTs and CN_xMWNTs, thus resulting in the creation of carbon nanoribbons and nanosheets. In panels (a-d) of Fig. 12.5, arrows guide the eye to point out the nanocutting of MWNTs and CN_xMWNTs.

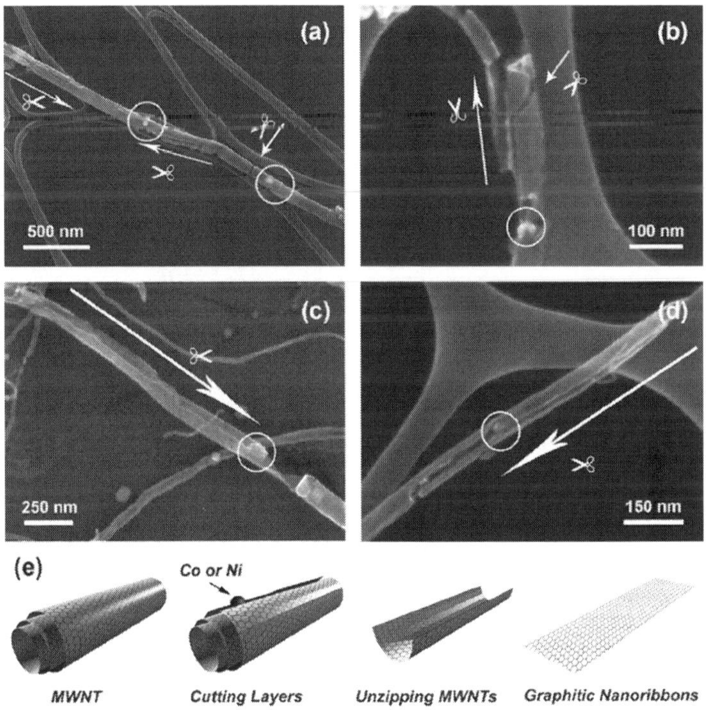

Figure 12.5 (a), (b), (d) SEM images of CN_xMWNTs cut with Ni nanoparticles. (c) SEM image of an MWNT on a Si substrate, with Co nanoparticles. Arrows guide the eye along the nanocutting and circles mark the presence of metal nanoparticles. (e) Schematic diagram representing different stages of cutting MWNTs along their axes with metal nanoparticles of Ni or Co. Reprinted with permission from Elías, A. L., Botello-Méndez, A. R., Meneses-Rodríguez, D., González, V. J., Ramírez-González, D., Ci, L., Muñoz-Sandoval, E, Ajayan, P. M., Terrones, H., and Terrones, M., Longitudinal cutting of pure and doped carbon nanotubes to form graphitic nanoribbons using metal clusters as nanoscalpels, *Nano Lett.*, **10**, pp. 366–372, 2010. Copyright (2010) American Chemical Society.

12.6 Current-Induced Breakdown

At very high electrical bias, MWNTs can undergo structural transformation through resistive heating and thermal bond breaking [254, 386, 387, 420]. This current-induced breakdown of nanotubes is exploitable for graphene nanoribbon production from MWNTs

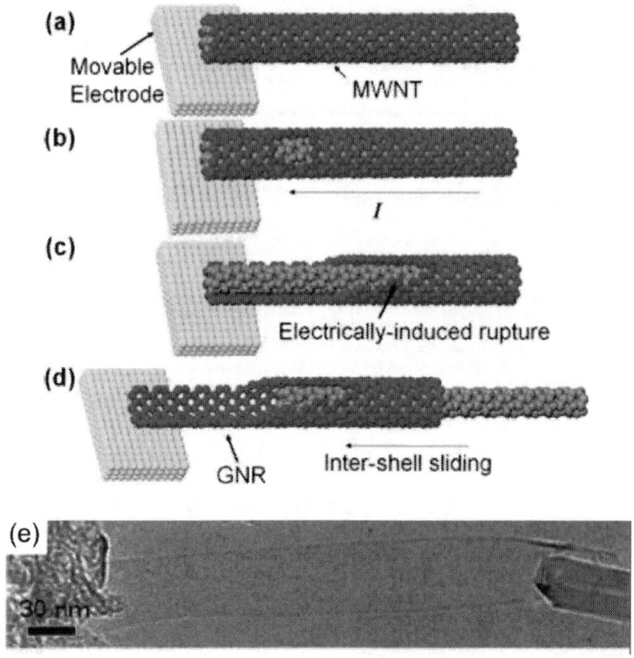

Figure 12.6 (a–d) Current-induced breakdown of MWNTs toward graphene nanoribbon fabrication. In the schematic, a double-wall carbon nanotube is chosen for simplicity: (a) MWNT before the partial wall rupture; (b) electrical current induces rupture of the outer wall of an MWNT; (c) partial outer-wall rupture of a MWNT results in a precursor nanoribbon, which is under the MWNT inner core; (d) intershell sliding between the nanoribbon and the inner core results in a suspended, electrically contacted nanoribbon. (e) TEM image of a nanoribbon derived from an MWNT via the electrical rupture and unwrapping method. The original MWNT from which the nanoribbon is derived is shown on the right side; it continues to serve as a mechanical support and electrical contact for the nanoribbon. Reprinted with permission from Kim, K., Sussman, A., and Zettl, A., Graphene nanoribbons obtained by electrically unwrapping carbon nanotubes, *ACS Nano*, **4**(3), pp. 1362–1366 2010. Copyright (2010) American Chemical Society.

[421]. The key is to promote thermally induced unwrapping of outer walls of MWNTs in a controlled way.

Figure 12.6 shows a pictorial representation of a fabrication process of extracting a nanoribbon from an MWNT. It is contacted to a movable electrode, and its outer walls are unwrapped via an applied electrical current through the contact and tube. With proper voltage bias control, only a part of the outer wall is severed (upper portion in Fig. 12.6 (b)) at which a local crack is generated. The crack triggers a subsequent erosion of the outer walls in the direction of current flow as shown in Fig. 12.6c; as a result, a precursor nanoribbon is developed that clings to the remaining MWNT inner core. Finally, the nanoribbon is removed from the eroded MWNT by inter-shell sliding (see Fig. 12.6(d)). The sliding process can be terminated when a desired amount of nanoribbon has been slid off. This leaves a preselected length of nanoribbon fully suspended in vacuum, with each end electrically and mechanically attached to a conducting electrode (the remaining inner core of the MWNT serves as one electrode).

The bottom of Fig. 12.6 shows a TEM image of a nanoribbon experimentally derived from the current-induced unwrapping technique. The original MWNT (30 nm diameter) from which the nanoribbon is derived is located on the right side of the nanoribbon. The length of the nanoribbon is about 300 nm and the width, uniform along the ribbon axis, is 45 nm, suggesting about half (circumferentially) of the MWNT outermost shells were vaporized during the electrical unwrapping process.

Chapter 13

Reinforcement Application

Out of association grows adhesion, and out of adhesion, amalgamation.

— Charlotte Brontë

13.1 As the Ultimate Reinforced Fibers

One reason for studying carbon nanotube mechanics is to explore the possibility of using them as reinforced fibers in composite materials [422]. Indeed, the exceptional mechanical properties of carbon nanotubes, combined with their low density ($\sim$1.3 g/cm^3), have continuously motivated the development of a new class of reinforced composite materials. In this chapter, we survey the theoretical background and experimental attempts of the nanotube-based reinforcement reported thus far.

From a historical perspective, the use of thin long fibers as a reinforcing agent dates back to thousands years ago. For instance, asbestos fibers[a] were used to reinforce clay posts about 3000 B.C.

[a]Asbestos is a set of six naturally occurring rock-forming minerals (i.e., silicate minerals). It is known to have high tensile strength, the ability to be woven, and resistance to heat and most chemicals.

Elastic and Plastic Deformation of Carbon Nanotubes
Hiroyuki Shima and Motohiro Sato

ISBN 978-981-4310-75-8 (Hardcover), 978-981-4364-15-7 (eBook)
www.panstanford.com

[423]. In the same century, Egyptians used straw to reinforce mud-bricks in establishing the Saqqara pyramid [424]. Nowadays, in industry, a variety of fine fibers made from alumina, glass, and silicon carbide are available for reinforcing macroscopic objects. These conventional fibers have typically tens micron diameters and millimeter lengths, for which the aspect ratio (length/diameter) is several tens more or less. The resultant fiber-reinforced composites find widespread use as structural materials in many fields such as aerospace, automotive, medical and military applications due to their superior strength-to-weight ratio, fatigue and fracture resistance and damping characteristics.

Despite the well-established usefulness of the traditional fibers, the ultimate reinforced fiber must be carbon nanotubes. We know that nanotubes have diameters ranging from 1 to 100 nm and lengths of up to millimeters [425]. Hence, they show extremely higher surface area per unit mass than traditional fibers, which implies much greater interaction with the embedding matrices. Furthermore, individual carbon nanotubes exhibit impressive mechanical properties; Young's modulus ($\sim$1 TPa) is superior to all conventional fibers, and the fracture strengths ($\sim$60 GPa) is an order of magnitude stronger than existing high-strength fibers. As a consequence of all the features mentioned above, the potential for nanotube-based reinforcement represents tremendous opportunity for application in the materials science and engineering.

For the accomplishment, however, there are still critical problems to overcome, a part of which will be explained in the next section.

13.2 Critical Challenges: Adhesion, Dispersion, Alignment

Toward the successful processing of nanotube-based reinforcement, one must fully understand the mechanical couplings between the embedded nanotubes and the surrounding matrix. This requirement is no different from that for conventional fiber-reinforced composites with macroscopic sizes [426]. However, the consequences of reducing the fiber's diameter into nanometric scales are less obvious. In the nanoscopic stage, interfacial coupling is through chemical interfacial bonding between individual atoms (covalent or non-covalent) across the interface [427], or it may be purely physical through non-bonded van der Waals (vdW) forces [428, 429]. In both cases, adhesion mechanism at the nanotube/matrix interface should differ far from macroscopic counterparts. Particularly in the vdW coupling situation, slipping between nanotubes and the matrix cause the reinforcement be deleterious. The issue of the interfacial coupling will be discussed in detail in Sections 13.6 and 13.7.

In addition to the interfacial-coupling problem, uniform dispersion of nanotubes within the matrix is an important challenge for efficient reinforcement [430]. The difficulty arises because nanotubes have a tendency to agglomerate[a] due to their large surface areas, high aspect ratios, and typically poor interactions with matrix [431]. The presence of agglomerates is extremely undesirable for reinforcement, as they act as defects leading to stress concentration and premature failure, specifically if the matrix does not fully penetrate the agglomerate during processing. On the other hand, a good dispersion makes nanotubes be loaded individually over a maximum interfacial area. In such preferred situation, each nanotube can contribute directly to toughening the surrounding matrix.

Even with improved dispersion and interfacial coupling, mechanical characterization of the resulting composites is still difficult because the distribution of the nanotubes is spatially random. Thus, alignment of the embedded nanotubes has been attempted in order to elucidate the reinforcement mechanisms. For instance, mechanical stretching was shown to provide alignment of nanotubes in composites [432], and the degree of alignment was examined by X-ray diffraction measurement. As another example, melt spinning of SWNTs in fiber form (with diameter $\sim$70 μm) was found to create a well-aligned nanotube composite [433]. If a perfect alignment is achieved, the enhancement in Young's modulus and fracture strength can be described by a simplified model as demonstrated in Sections 13.3 and 13.5.

In summary, several key challenges remain to be met for successful nanotube-based reinforcement [434]. First, nanotubes

[a] In particular, SWNTs tend to agglomerate into "bundles" that consist of many parallel nanotubes loosely bound to each other by weak vdW forces.

with intrinsically good mechanical properties must be obtained in reasonable quantity at acceptable cost. Second, the nanotubes must be processed in such a way of a homogeneous dispersion and an appropriate degree of interfacial coupling. Finally, well-aligned nanotube configuration in the embedding matrix is favored. These overall requirements are common to all kinds of Nanotube composites ranging from polymers to ceramics and metal matrices discussed in Sections 13.8 and 13.9.

13.3 Mixture Rules and Beyond

First of all, we are interested in what amount of reinforcement can be achieved in general cases by inserting high-strength fibers into a host matrix. The question can be answered in part by modeling the mechanical properties of fiber-reinforced composites. The modeling has been carried out since the 1950s [435], long before the boost of carbon nanotube research triggered in 1991 [1]. A number of sophisticated models have been suggested thus far, and a part of them can apply to nanotube-based composites. Among them, we begin with the simplest and most common one, called "the rule of mixtures" or the "mixture rule" [436, 437].

13.3.1 *Mixture Rule: Long-Fiber Situation*

Suppose an isotropic, elastic matrix is filled with perfectly aligned elastic fibers. In deriving the mixture rule for Young's modulus, we assume a perfect interfacial coupling over the whole nanotube surface with no slippage between the fibers and matrix (Fig. 13.1(a)). The strain of the matrix is transferred completely to the fiber; this ensures continuity of displacement at each point of the interface. We also assume that all fibers span the full length of the matrix. Hence, subjected to a stress in the fiber alignment direction, the matrix and fibers will be equally strained. Under these conditions, the composite's Young's modulus, $Y_{\rm C}$, in the alignment direction is given by

$$Y_{\rm C} = Y_{\rm f} \cdot v_{\rm f} + Y_{\rm m} \cdot (1 - v_{\rm f}), \tag{13.1}$$

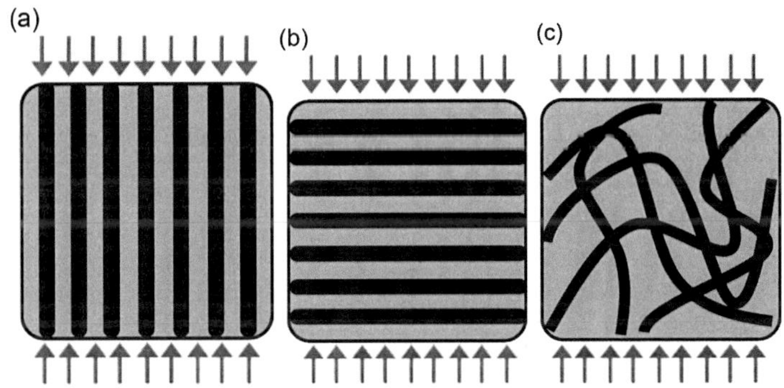

Figure 13.1 Schematics of different materials filled with long fibers under compression. Solid thick curves are the fibers, hatched regions represent the host matrix. (a) Longitudinally loaded composite showing the modulus given by Eq. (13.1). (b) Transversely loaded composite corresponding to Eq. (13.2). (c) Randomly dispersed composite. Reprinted with permission from Ci, L., Suhr, J., Pushparaj, V., Zhang, X., and Ajayan, P. M., Continuous carbon nanotube reinforced composites, *Nano Lett.*, **8**(9), 2762–2766, 2008. Copyright (2008) American Chemical Society.

where $Y_{\rm f}$ and $Y_{\rm m}$ are Young's moduli of fibers and the matrix, respectively; $v_{\rm f}(< 1)$ denotes the fiber volume fraction. Equation (13.1) is the rule of mixtures describing Young's modulus enhancement by applying the fiber reinforcement [436, 437]. It is easy to derive Eq. (13.1) from our assumptions as to the strain, $\varepsilon_{\rm f} = \varepsilon_{\rm m} = \varepsilon_{\rm C}$, and the obvious relation, $\sigma_{\rm C} = v_{\rm f}\sigma_{\rm f} + (1 - v_{\rm f})\sigma_{\rm m}$, satisfied by the stress expressed by $\sigma_i = Y_i\varepsilon_i$ $(i = f, m, C)$. Here, the subscripts f, m, C indicate the fiber, matrix, and composite, respectively.

The above discussion is based on the assumption that the stress is applied to the axial direction of fibers. A similar argument is possible by supposing the stress applied to the direction perpendicular to the fiber's axis (Fig. 13.1(b)). If all fibers are sufficiently long, the composite's Young's modulus for such perpendicular situation reads

$$Y_{\rm C} = \frac{Y_{\rm m}Y_{\rm f}}{Y_{\rm m}v_{\rm f} + Y_{\rm f}(1 - v_{\rm f})}. \tag{13.2}$$

Equation (13.2) is derived from the following relations: $\sigma_{\rm f} = \sigma_{\rm m} = \sigma_{\rm C}$, $\sigma_i = Y_i\varepsilon_i$ $(i = f, m, C)$, and $\varepsilon_{\rm C} = v_{\rm f}\varepsilon_{\rm f} + (1 - v_{\rm f})\varepsilon_{\rm m}$. The final formula is referred to the fact that the fiber's fraction in the stress direction per unit length equals to $v_{\rm f}$.

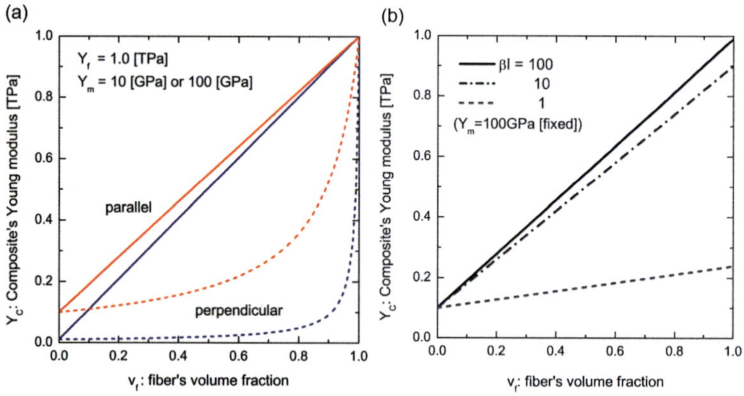

Figure 13.2 Composite's Young's modulus Y_C as a function of the fiber volume fraction v_f. (a) Data based on the mixture rule for the parallel (solid line) and perpendicular (dashed) situations, given by Eqs. (13.1) and (13.2), respectively. (b) Data based on the shear-lag model represented by Eq. (13.3).

Figure 13.2(a) shows the difference in the v_f-dependence of Y_C between Eqs. (13.1) and (13.2), i.e., the parallel and perpendicular situations. The Y_C in parallel cases increases linearly with v_f, from $Y_C = Y_m$ at $v_f = 0$ to $Y_C = Y_f$ at $v_f = 1$. On the other hand, the Y_C in perpendicular cases is convex downward, coinciding with the parallel counterpart both at $v_f = 0$ and $v_f = 1$. It should be noted that each curve serves as the lower and upper bound, respectively, of the enhanced modulus for sufficiently long fiber cases; an intermediate value for Y_C will be obtained in practice, since fibers can disperse randomly in a host matrix as presented in Fig. 13.1(c).

13.3.2 *Shear-Lag Model: Short-Fiber Situation*

While Eqs. (13.1) and (13.2) give a general trend, they are valid only in an idealized situation such that the long fibers run continuously through the matrix. Quite often, fibers are much shorter than the dimension of an embedding matrix. The composite modulus Y_C for such short-fiber cases deviates from the mixture rules mentioned above, and can be described by the shear-lag model (or called Cox's model) [438].

The shear-lag model postulates that short fibers with length ℓ are distributed randomly in positions but they all are perfectly parallel to the stress direction. Then, the composite modulus Y_C is given by [439]

$$Y_C = Y_f \cdot v_f \left[1 - \frac{\tanh(\beta\ell/2)}{\beta\ell/2}\right] + Y_m \cdot (1 - v_f). \tag{13.3}$$

Here, $\beta = \sqrt{c/Y_f}$ with a constant c whose values depends mainly on the spatial configuration of fibers in the matrix. Provided all fibers have a diameter $2r_0$ and they are separated in the mean by $2R$ (between the long axes of the neighboring fibers), we obtain

$$c = \frac{2\tau_m}{r_0^2 \ln(R/r_0)} \quad \text{and thus} \quad \beta = \frac{1}{r_0}\sqrt{\frac{2\tau_m}{Y_f \cdot \ln(R/r_0)}}, \tag{13.4}$$

where τ_m is the matrix's shear modulus. For a given r_0 (and R), therefore, Eq. (13.3) reduces to the mixture rule in the limit of $\ell \to \infty$ since $(\tanh x)/x$ converges to zero if $x \to \infty$.

Figure 13.2(b) illustrates the result shown in Eq. (13.3) for several different values of $\beta\ell$. When $\beta\ell = 100$ or more, the behavior of Y_C is almost same as in the long-fiber situation represented by Eq. (13.1). But if $\beta\ell \sim 1$ or less, no significant reinforcement can be obtained with increasing v_f. For a rough estimate of $\beta\ell$ for actual nanotube-based composites, we suppose $Y_f = 1$ TPa, $\tau_m = 50$ MPa (a typical value for polymers), $R = 10$ nm and $r_0 = 1$ nm. Then we have $\beta = 10\ \mu\text{m}^{-1}$, which means that the nanotube's length $\ell = 10\ \mu$m leads to $\beta\ell = 100$ that is enough to attain the efficient reinforcement.

The derivation of Eq. (13.3) is outlined below [438, 439]. Denote by $u(x)$ the deformation amplitude of fiber's volume element at a point x in the axial direction, and by $v(x)$ the matrix's counterpart at the same point that would be observe if the fibers were absent from in the host matrix. To make an analytic argument, it is assumed that the axial stress σ_f applied to the fiber at x is given by $d\sigma_f/dx = c[u(x) - v(x)]$ with a constant c. Keeping in mind Hooke's law $\sigma_f = Y_f(du/dx)$ and the composite's strain $dv/dx = \varepsilon_C \equiv$ const, we obtain the differential equation

$$\frac{d^2\sigma_{\mathrm{f}}}{dx^2} = c\left(\frac{\sigma_{\mathrm{f}}}{Y_{\mathrm{f}}} - \varepsilon_{\mathrm{C}}\right), \tag{13.5}$$

whose solution under the boundary conditions $\sigma_{\mathrm{f}}(0) = \sigma_{\mathrm{f}}(\ell) = 0$ is given by

$$\sigma_{\mathrm{f}}(x) = Y_{\mathrm{f}}\varepsilon_{\mathrm{C}}\left[1 - \frac{\cosh\beta(\ell/2 - x)}{\cosh(\beta\ell/2)}\right], \quad (0 \leq x \leq \ell/2) \tag{13.6}$$

where $\beta = \sqrt{c/Y_{\mathrm{f}}}$. As a result, the averaged one $\overline{\sigma_{\mathrm{f}}} \equiv (2/\ell)\int_0^{\ell/2}\sigma_{\mathrm{f}}(x)dx$ equals to

$$\overline{\sigma_{\mathrm{f}}} = Y_{\mathrm{f}}\varepsilon_{\mathrm{C}}\left[1 - \frac{\tanh(\beta\ell/2)}{\beta\ell/2}\right], \tag{13.7}$$

from which we obtain Eq. (13.3).

13.4 Reinforcement by Millimeter-Long Nanotubes

Until recently, limited lengths of carbon nanotubes (up to a few hundreds micrometers) can be obtained in synthesis; this implies that nanotubes are considered short fibers, randomly dispersing in the matrix of macroscopic sizes. But we learned in Section 13.3.2 that short fibers are not the best for high-strength structural composites. Hence, it is interesting to fabricate composites with continuous nanotubes threading over the composite matrix. Indeed, the recent optimization in aligned nanotube growth allows to obtain nanotubes of millimeters lengths [440, 441]; this provides an opportunity for the proper utilization of the extreme high modulus and strength predicted for nanotubes in structural composites.

Figure 13.3 shows millimeter-long aligned multi-walled carbon nanotube arrays grown on solid substrates ($\sim$3.5 mm high over an area of 2.5 $\times$ 2.5 mm^2) [442]. In the as-grown nanotube arrays, only 5% in volume is occupied by nanotubes, and the rest is empty. The composites were prepared by infiltrating a polymer into the interstitial spaces of the nanotubes. As a result, the intrinsic high porosity of the nanotube arrays is completely replaced by the polymer matrix, and the composite can be characterized as

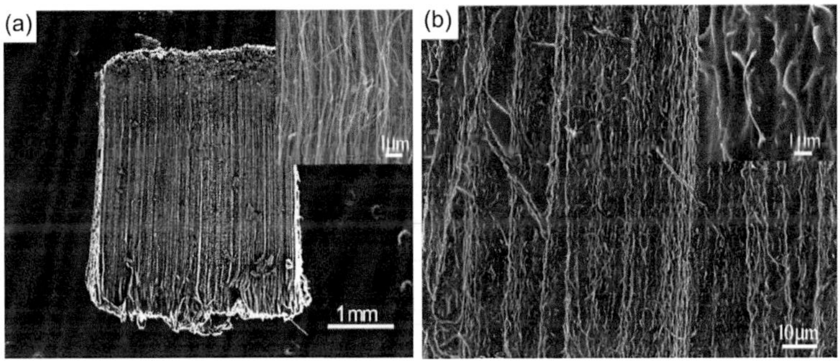

Figure 13.3 (a) Millimeter-long vertically carbon nanotube arrays. The excellent vertical alignment of the nanotubes is observed, and high porosity is also seen at a higher magnification (inset), (b) Continuously reinforced carbon nanotubes-polymer composite. The polymer chains are seamlessly infiltrated into open space between nanotubes. Reprinted with permission from Ci, L., Suhr, J., Pushparaj, V., Zhang, X., and Ajayan, P. M., Continuous carbon nanotube reinforced composites, *Nano Lett.*, **8**(9), 2762–2766, 2008. Copyright (2008) American Chemical Society.

a continuous nanotube array composite with about 5% volume fraction of nanotubes.

Figure 13.4 shows compressive stress–strain characteristics for different materials (pure nanotube array, pure polymer, and composites) with almost identical dimension [442]. The composite under longitudinal compression along the nanotube orientation (longitudinal composite) exhibits dramatic increase in stiffness more than 600% over the entire strain range. The enhanced longitudinal modulus is because the polymer chains around the nanotubes make nanotube's buckling far more difficult [443–445]; such enhancement cannot be expected in short-fiber situations, thus being manifestation of the advantages of the sufficiently long-fiber case.

For comparison, transversely or randomly oriented nanotube–polymer composites with the same volume fraction (5%) were also fabricated and tested. It follows from Fig. 13.4 that the longitudinal composite is far more effective in reinforcement than other two composites, as in agreement with the theoretical discussion in Section 13.3.1.

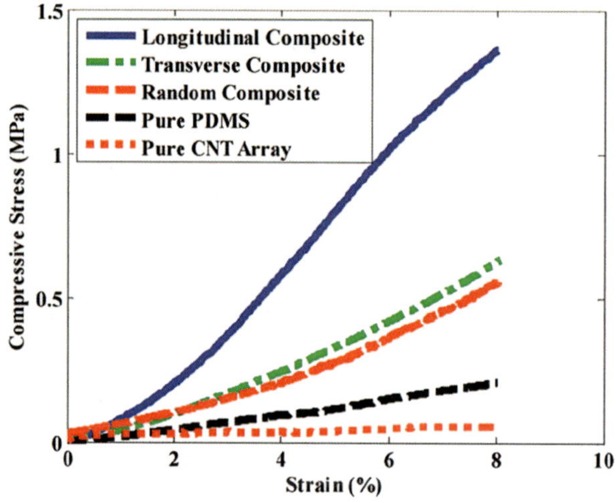

Figure 13.4 Compressive stress–strain curves for composites, pure polymer, and pure nanotube array. The most significant enhancement in Young's modulus (= the slope of the curve) is obtained when the compressive stress is applied in the longitudinal direction (i.e., parallel to the nanotube array). Reprinted with permission from Ci, L., Suhr, J., Pushparaj, V., Zhang, X., and Ajayan, P. M., Continuous carbon nanotube reinforced composites, *Nano Lett.*, **8**(9), 2762–2766, 2008. Copyright (2008) American Chemical Society.

13.5 Critical Length for Fiber Breaking

Next we consider the fracture strength of the reinforced composites. For sufficiently long fibers (i.e., they span the whole dimension of the host matrix), the mixture rule similar to Eq. (13.1) basically holds for the composite strength σ_C^{max} such as

$$\sigma_C^{max} = \sigma_f^{max} \cdot v_f + \sigma_m^{max} \cdot (1 - v_f). \tag{13.8}$$

Here, σ_f^{max} and σ_m^{max} are the fracture strengths of fibers and matrix, respectively.

When the fibers have finite length ℓ shorter than the matrix dimension, Equation (13.8) should be modified by taking into account the surpression of stress transfer between the fiber and matrix. The stress transfered increases with a distance from the end of the fiber along the fiber's axis (see Fig. 13.5). The maximum stress arises at the midpoint ($x = \ell/2$) of the fiber. Therefore, if the maximum exceeds the fiber's fracture strength σ_f^{max}, the applied load to the matrix may cause a fiber breaking before the fracture of the embedding matrix itself. In this context, we can define the critical length [439], ℓ_c, by supposing that maximum stress arising at $x = \ell_c/2$ equals to σ_f^{max}. When $\ell < \ell_c$, enough stress cannot be transferred to the fibers to break them. As a result, the matrix fails with the fiber pulling out of the matrix, where the reinforcing ability of the fibers is not fully manifested. In other words, short fibers ($\ell < \ell_c$) carry load less efficiently so that they can be thought of as having a lower effective modulus for reinforcement purposes. For very long aligned fibers ($\ell \gg \ell_c$), in contrast, the fibers can assume significant amount of stress and sustain until they break under large applied stress just above the fiber's fracture strength σ_f^{max}.

It can be derived that the critical fiber length ℓ_c is represented by [439]

$$\ell_c = \frac{r_0}{\tau_y}\sigma_f^{max}, \tag{13.9}$$

where τ_y is the interfacial shear yield stress between fibers and matrix. Equivalently, it is rewritten in the form of the aspect ratio,

$$\lambda_c \equiv \frac{\ell_c}{r_0} = \frac{\sigma_f^{max}}{\tau_y}. \tag{13.10}$$

Now let us estimate the realistic value of ℓ_c (or equivalently λ_c) for actual nanotube-polymer composites. Recent experiments on nanotube bundles [29, 30, 216] have reported the fracture strength of the nanotubes to be in the range of 50 GPs. With a typical value of 50 MPa for τ_y at the nanotube-polymer interface, the limiting value of the aspect ratio is 10^3. Therefore, for an optimum load transfer with a nanotube of 10 nm diameter, for instance, the nanotube should be at least 10 μm long, which is readily feasible in the actual nanotube synthesis.

13.6 Interfacial Coupling Strength

13.6.1 *Breakthrough Wanted*

It has been broadly accepted that the interface between the nanofiber and the matrix dictates the overall properties of the

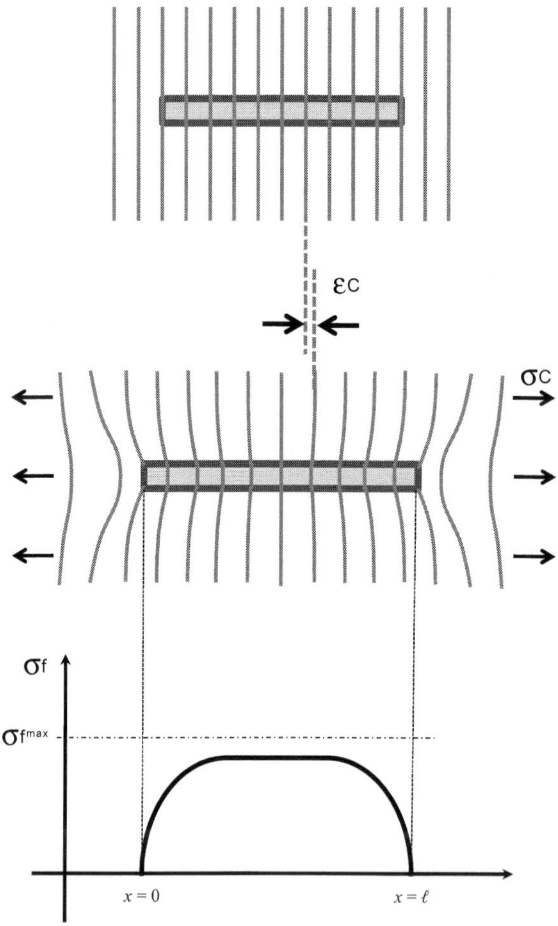

Figure 13.5 Illustration of the stress transfer to the fiber (horizontal rod) from the embedding matrix subjected to the stress σ_C. The vertical stripe represents the deformation profile of the matrix (and the fiber), from which the matrix's strain ϵ_C is deduced. The bottom plot shows the stress distribution within the fiber, in which the stress takes the maximum at the mid-point $x = \ell/2$. When the maximum value exceeds the fiber's fracture strength σ_f^{max}, then the fiber fails.

nanocomposite. Still, improving the interfacial strength has remained to be a challenging task for efficient reinforcement [427, 428, 446–453, 478]. Despite more than one decade of research, the ability to manipulate the nanotube/matrix interface to achieve desired mechanical properties still remains rather limited and largely empirical. The difficulties can be traced to the following two fundamental issues.

The first is the absence of a reliable testing platform at the nanoscale that enables reproducible measurements of interfacial strength. While various techniques are availiable for traditional macroscopic composites to quantity the fiber–matrix interface [454–460], the extension of these techniques to nanotube–polymer systems is exceedingly difficult due to the small size of the nanotubes. The second issue to be noted is the lack of a theoretical framework that connects nanoscale interfacial features to macroscopic properties. For instance, a pullout test (one of the standard methods measuring the interfacial coupling in a macroscopic composite) provides only an spatially averaged value of the interfacial shear strength at the time of failure. But in nanoscale, the magnitude of interfacial stress fluctuate over the interface region. Toward better understanding of the debonding process at the interface, therefore, more sophisticated theories to predict the stress distribution in composites during inclusion pullout should be anticipated [459, 460].

Against the background noted above, the reader must bear in mind that the two issues (experimental and theoretical) are to be addressed in future study toward the successful synthesis of nanotube-reinforced composites.

13.6.2 *Category of Interfacial Couplings*

Following are the three classes of interaction between a nanotube and a polymer matrix [422]:

(i) chemical bonding
(ii) nanomechanical interlocking
(iii) non-bond interactions (van der Waals and electrostatic)

The reader may suspect that strong chemical bonds with a surrounding matrix, raised as (i) in the above list, should be forbidden at the nanotube surface because of the perfect sp^2-hybridized structure of the seamless carbon walls. But this is somewhat up to the type of matrix used, and in fact, a few polymers turned out to chemically bond with pristine nanotubes [428, 461]. Quantum mechanical simulations have also supported the possibility of such chemical bonding [462]. Particularly interesting is the influence of nanotube functionalization on the interface, where defects are intentionally introduced on the nanotube surface to provide sites for covalent bonding with the matrix [463–466]. Several examples will be accounted for in Section 13.8.4.

Nanomechanical interlocking also appears to be difficult in nanotube composites due to the atomically smooth surface of a carbon nanotube. Nevertheless, actual nanotubes involve local non-uniformities such as kinks, bends, and changes in diameter. Micromechanical interlocking occurs at these local non-uniformities [467]. The ideal situation for the occurrence is that: nanotubes are strong enough and inter-connected or long enough to block the movement of the polymer chains [468, 469].

As to non-bond interactions, controversial results are yet to be fully solved. Molecular-dynamics (MD) simulations [428] showed that non-bond interactions yield the interfacial strength as high as 100 MPa, which is one order of magnitude higher than typically reported conventional (macroscopic) fiber–polymer composites [470]. The results also can be compared to an experimentally determined interfacial shear strength of 500 MPa estimated for nanotubes in a polymer matrix [470], in which stress-induced fragmentation method was employed. On the other hand, a shear lag model–based estimation [427] has revealed that van der Waals forces give a very low interfacial strength (2 MPa) for a nanotube–polymer composite. Another MD-based study [471] revealed that the non-bond binding energy and friction forces play only a minor role in determining the interfacial strength. Instead, in the absence of chemical bonding, helical conformation of polymers around nanotubes could facilitate the creation of a strong interface. Put in other words, the interfacial strength results from molecular-level entanglement of the two phases. Keeping in mind the series of controversial results, continued efforts will be anticipated to quantify the interfacial couplings based on the non-bond interaction.

13.7 Pullout Test

For macroscopic composites, several techniques are available for the measurement of interfacial strength, such as single-fiber pullout [472] or push-out [473], fragmentation [474], and Raman spectroscopy [475]. Among them, single-fiber pullout is a standard and the preferred technique for interface characterization, because it directly estimates the interfacial strength [450, 476]. In a pullout test, we measure the force required to pull out an individual fiber completely from the composite and then determine the interfacial strength using the fiber geometry and measured force values. However, the technique is challenging to adapt for nanotubes. The nanometric size makes it difficult to ensure firm and reliable gripping of the nanotubes during the pullout experiment and requires high force and displacement resolution for meaningful quantitative characterization.

The pioneering pullout test was performed in 2002 using a scanning probe microscope (SPM) tip [478]. The interfacial interaction was determined by detaching individual SWNT bundles and MWNTs from the host matrix. After dragging across the hole with the SPM tip, the MWNT had been fully pulled out from the top of the hole and partially drawn out of the bottom of the hole leaving a cylindrical cavity, as a result of loading the nanotube at its center (analogous to a cable with two fixed ends and a center loading). The nanotube was bent in the direction of the pull. The shear strength for the MWNT was calculated by dividing this pullout force by the interfacial area of the embedded nanotube. The values estimated ranges approximately from 30 to 400 MPa, which implies that relatively strong interaction at the interface are in principle attainable. The subsequent experimental work done by the same research group [450] reported a nanoscale pullout test using AFM on an MWNT–polymer matrix, suggesting an interfacial shear stress of 47 MPa (see Fig. 13.6).

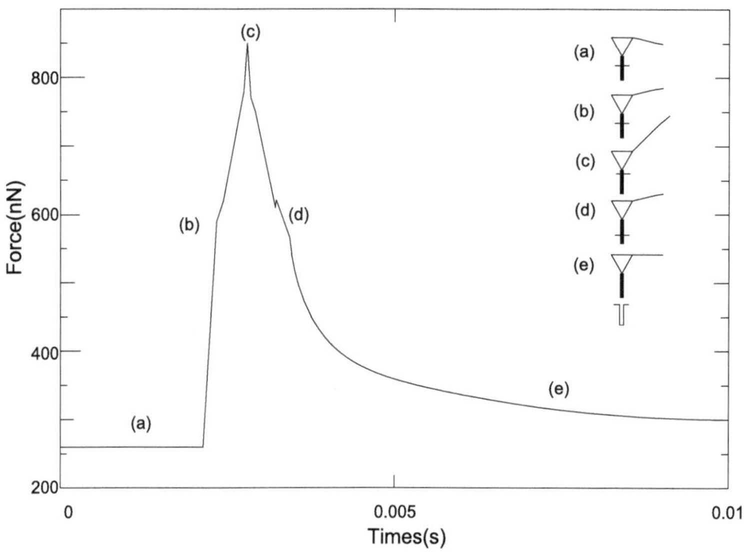

Figure 13.6 Typical plot of pullout force (taken from the AFM cantilever deflection) against pullout time. At (a), the nanotube is embedded in the polymer. As the nanotube is pulled away from the polymer, the cantilever bends away (b) until the maximum force, corresponding to the maximum cantilever bending deflection, is achieved (c). Pullout then occurs (d), resulting in the eventual complete separation of the nanotube from the polymer (e). Adapted from Ref. [450].

The two experiments mentioned above are very difficult to perform and the scatters in the data are significantly high. The obstacles in obtaining high-quality, quantifiable data stems from a number of factors: the difficulty in determining the embedded length of the nanotube within the polymer; the limited ability to align the embedded nanotubes perpendicular to the fracture surface; the imperfect straightness of the embedded nanotubes; and the ill-application of a pure tensile load to the embedded nanotube. Continued development and refinement of nanoscale pullout tests will provide a clue to eliminate these obstacles from the interpretation of our pullout experiments.

13.8 Polymer Reinforcement

Various fabrication techniques have been developed thus far toward nanotube–polymer composites with significant reinforcement [479]. This section is devoted to give a brief review of the familiar methods to synthesize the composites and the resultant improvement in mechanical properties.

13.8.1 *Solution Process*

The most straightforward approach for preparing nanotube–polymer composites is to mix the nanotubes and polymer in a suitable solvent; then the solvent is evaporated to form a composite film. In practice, the choice of solvent is often made based on the solubility of the polymer. But we should be aware of pristine nanotubes cannot be well dispersed in most solvents; hence, an additive such as a surfactant is commonly used to disperse the nanotubes before mixing with the polymer solution [480–482].

The first synthesis of the solution-based composites was reported in 1999 [483]. The resulting composites involve the nanotubes with weight fractions of up to 60%, showing only a slight reinforcement with Young's modulus enhancement from 6 GPa for the polymer to 12 GPa for the composite. Subsequent efforts for the improvement have enabled to give more significant reinforcement with modulus enhancement (from 1.92 GPa to 7.04 GPa) at 0.6% nanotube volume fraction, and strength enhancement (from 81 MPa to 348 MPa) as demonstrated in [484].

13.8.2 *Melting Process*

Obviously, the solution-based approach explained in Section 13.8.1 is not suitable for many classes of polymers that are insoluble. Melt processing is a common alternative, which is particularly useful for dealing with thermoplastic polymers.[a] Merits of this technique

[a] A thermoplastic polymer is a class of polymers that melts when heated and freezes when cooled. The binding forces between molecules are the results of vdW forces and mechanical entanglement between the polymer chains. The bonds can be easily broken upon heating, which is why thermoplastic materials turn to a liquid.

are its speed and simplicity, not to mention its compatibility with standard industrial techniques [485, 486].

An early study on the melt mixing [487] in 2001 was done through a polymer in a moulder at 200°C. The melt involving nanotubes was then compression moulded (under 8–9 MPa at 210°C) in a hydraulic press to give composite slabs. The composite exhibited an increase in Young's modulus from 0.7 to 1.63 GPa at 17 wt.% nanotubes. Microscope studies showed that dispersion was good even at high nanotube concentration.

Afterward, impressive reinforcement was successful [488, 489]: a threefold increase in Young's modulus from 0.4 to 1.24 GPa on addition of only 2 wt.% nanotubes and and a significant increase in yield strength from 18 to 47 MPa with similar increases in fracture strength. These impressive results were attributed to very good dispersion and adhesion as observed by microscopy measurements.

13.8.3 *Thermosetting Process*

Because of their wide range of industrial uses, thermosetting polymers[a] (epoxy resins, in particular) have been widely studied as a potential matrix for nanotube-based composites. Thermosetting polymers are generally hardened when heated or mixed with a catalyzing agent. In most cases, they begin their lives in liquid form, facilitating nanotube dispersion. Hardening is then carried out to convert the liquid composite to the final solid state.

The first record of the formation of epoxy–nanotube composites was presented in 1994 [477], where nanotubes were aligned within the epoxy matrix by the shear forces induced by cutting with a diamond knife. However, no quantitative mechanical measurements were made.

The first true mechanical study was made in 1998 [446], in which the stress–strain properties of a nanotube–epoxy composite in both tension and compression were presented. In tension, no significant increase in the modulus was observed. Nevertheless, better results

[a]A thermosetting polymer is one that does not melt down when heated (unlike thermoplastic polymers) but forms a rigid, cross-linking network of polymer chains through heat treatment or chemical reaction.

were seen in compression with an increase in modulus from 3.63 to 4.5 GPa. The difference between tension and compression was explained by the fact that load transfer in compression can be thought of as a hydrostatic pressure effect while load transfer in tension relies on the coupling at nanotube/polymer interface.[a] Another study in 2002 has confirmed a significant increase in modulus from 4.2 to 5 GPa at only 0.1 wt.% [491]. Interestingly, according to the rule of mixtures, this value is close to the maximum possible reinforcement for a material filled with perfectly aligned, well-dispersed nanotubes displaying perfect stress transfer. Around the same time, successful reinforcement exhibiting good all-round mechanical properties were reported such as a doubling in the fracture strength [492] and that in Young's modulus [493].

13.8.4 *Functionalization Process*

It is widely known that due to the smooth graphene-like surface of nanotubes, there is a lack of interfacial bonding between polymer matrix and carbon nanotubes. Chemical functionalization of the nanotube surface is expected to solve the problem. The main advantage of the functionalization method is the capability of joining polymer chains onto the nanotube surface. Covalently bonded long-chain molecules entangle with the surrounding polymer matrix, as a result of which a very strong interfacial interaction is created. In addition, the functional groups attached to the nanotube surface act to make the nanotubes more compatible both with polymer hosts and solvents. This tends to dramatically improve the nanotube dispersion and hence further improve composite properties. Notably, the functionalization technique is important for the preparation of insoluble and thermally unstable polymers, which cannot be processed by the three previous methods described in Sections 13.8.1, 13.8.2 and 13.8.3.

Two main strategies for the covalent joining of polymers to the nanotubes have been reported: "grafting from" [494] and "grafting

---

[a] It should be pointed out that later studies showed the reverse to be true, i.e., load transfer in tension gives a positive result but none in compression [490].

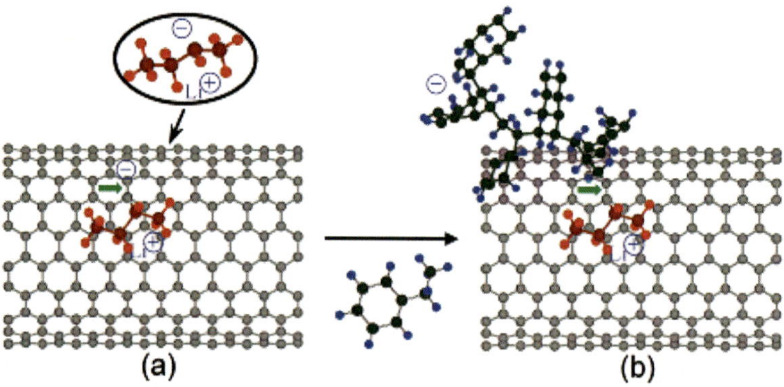

Figure 13.7 Schematic of the "grafting from" process. (a) An initiator element is assigned on the nanotube surface. (b) Successive addition of negatively charged monomers results in the formation of a living polymer chain. Reprinted with permission from Viswanathan, G., Chakrapani, N., Yang, H., Wei, B., Chung, H., Cho, K., Ryu, C. Y., and Ajayan, P. M., Single-step in situ synthesis of polymer-grafted single-wall nanotube composites, *J. Am. Chem. Soc.*, **125**(31), 9258–9259, 2003. Copyright (2003) American Chemical Society.

to" [495]. The "grafting from" method is based on the initial immobilization of initiators onto the nanotube surface followed by polymerization of appropriate monomers with the formation of the polymer molecules bound to the nanotube (see Fig. 13.7). The advantage of this technique is that composites with quite high grafting density can be prepared. However, this method requires a strict control of amounts of initiator and substrate as well as a control of conditions for polymerization reaction. The "grafting to" approach is based on attachment of already preformed polymer molecules to functional groups on the nanotube surface via chemical reactions. An advantage of this method is that preformed commercial polymers of controlled mass and distribution can be used. The main limitation of the "grafting to" technique is that initial binding of polymer chains hinder the approach of additional macromolecules to the surface, leading to low grafting density. Also only polymers containing reactive functional groups at the end of the chain can be used.

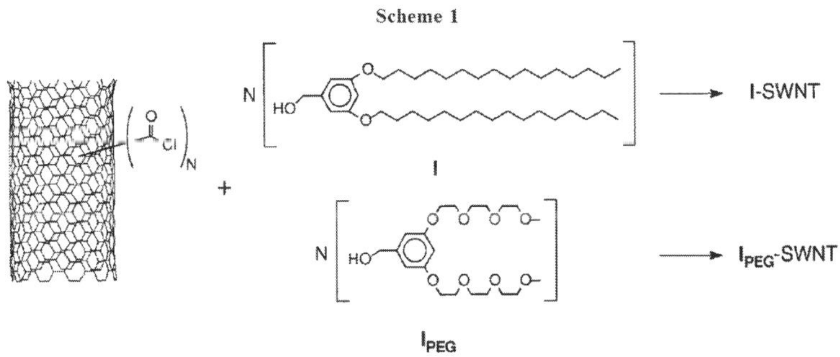

Figure 13.8 Illustration of the "grafting to" process. Already-prepared long polymer chains are attached to the nanotube surface via chemical reaction. Reprinted with permission from Fu, K., Huang, W., Lin, Y., Riddle, L. A., Carroll, D. L., and Sun, Y. P., Defunctionalization of functionalized carbon nanotubes, *Nano Lett.*, **1**(8), 439–441, 2001. Copyright (2001) American Chemical Society.

13.9 Ceramics and Metal Reinforcement

The vast majority of nanotube composite research has focused on polymer matrices, while comparatively few investigations have explored inorganic (ceramic or glass) matrices and the potential toughening mechanisms [496, 497]. The ceramic use allows us to overcome some drawbacks of polymers such as aging and limited temperature range of applications. Most studies on nanotube–ceramic composites ultimately aim to increase the fracture toughness,[a] instead of fracture strength or Young's modulus. This is because of the brittle nature of ceramic; in fact, though ceramics have high stiffness with relatively low density, their brittleness impedes the use as structural materials. Potentially, nanotubes might be desirable as a reinforcement for ceramics because of their exceptional resilience, leading to exceptional toughness and creep resistance.

[a]Fracture toughness is an indication of the amount of stress required to propagate a pre-existing crack (or other kinds of flaw) in a given matrix. It describes the ability of a material containing a crack to resist fracture, thus being a distinct quantity from the fracture strength. Since the presence of cracks is not completely avoidable in the real processing, it serves as one of the most important properties of any material for virtually all design applications.

The primary ceramic material attempted is aluminum oxide (or simply "alumina") signified by Al_2O_3 [498–506]. It was found that [499] the fracture toughness of nanotube–alumina composites becomes three times higher than that of pure alumina crystals by introducing 10 wt% of nanotubes. The yield strength[a] and the ultimate strength[b] are also enhanced by almost a twofold with an addition of 2 wt% of nanotubes [506]. Besides, nanotube–silicon carbide (SiC) composites with 10 wt.% nanotubes showed a 10% improvement in the strength and fracture toughness compared with the raw ceramics. These modest improvements were attributed to nanotube/matrix debonding and crack deflection. Other efforts [507–511] have made to synthesize carbon nanotubes *in situ* to form nanotube–metal oxide composites. However, the incorporation of the long nanotube bundles did not provide the expected improvement in mechanical properties.

It is interesting to mention that in the past few years, the metal–nanotube composites synthesis has developed intensively [512]. Most of them focused of Al-nanotube composites [506, 513–515], while the rest of the studies were devoted to Cu- [516–518], Fe- [519], or Co-based [520] composites. Overall, however, research on the mechanical performance of nanotube–inorganic matrix composites (including metals and ceramics) is at a relatively early stage. The reports of modest improvements in mechanical properties do not, for the most part, provide clear evidence linking the quantitative performance data to the actual mechanisms involved. Further work relating properties to mechanism is clearly required.

---

[a] Yield strength is the maximum stress below which elastic deformation can be measured. Beyond the yield strength, therefore, we will observe plastic deformation. From the definition, a manual load–unload procedure is required to determine the yield strength of a given material.

[b] Ultimate strength, often called tensile strength or ultimate tensile strength, is the maximum stress that a material can withstand while being stretched or pulled before necking, at which the specimen's cross section starts to significantly contract.

Bibliography

1. Iijima, S. (1991). Helical microtubules of graphitic carbon, *Nature*, **354**, pp. 56–58.
2. Osawa, E. (1970). Superaromaticity, *Kagaku*, **25**, pp. 854–863. (in Japanese).
3. Kroto, H. W. Heath, J. R., O'Brien, S. C., Curl, R. F., and Smalley, R. E. (1985). C_{60}: Buckminsterfullerene, *Nature*, **318**, pp. 162–163.
4. Zhao, J., and Zhu, J. (2011). Electron microscopy and *in situ* testing of mechanical deformation of carbon nanotubes, *Micron*, **42**, pp. 663–679.
5. Radushkevich, L. V., and Lukyanovich, V. M. (1952). O strukture ugleroda, obrazujucegosja pri termiceskom razlozenii okisi ugleroda na zeleznom kontakte, *Zurn. Fisic. Chim.*, **26**, pp. 88–95. [Translated into *Russ. J. Phys. Chem.*, **26**, pp. 88–95 (1952)].
6. Oberlin, A., Endo, M., and Koyama, T. (1976). Filamentous growth of carbon through benzene decomposition, *J. Cryst. Growth.*, **32**, pp. 335–349.
7. Boehm, H. P. (1997). The first observation of carbon nanotubes, *Carbon*, **35**, pp. 581–584.
8. Monthioux, M., and Kuznetsov, V. L. (2006). Who should be given the credit for the discovery of carbon nanotubes? *Carbon*, **44**, pp. 1621–1623.
9. The talk was presented in 1979, but the relevant paper was published in 1999 as: Abrahamson, J., Wiles, P. G., and Rhoades, B. L. (1999). Structure of Carbon Fibers Found on Carbon Arc Anodes *Carbon*, **37** pp. 1873–1874.
10. US 4663230, Tennent, Howard G., Carbon fibrils, method for producing same and compositions containing same, issued 1987-05-05.
11. Poncharal, P., Wang, Z. L., Ugarte, D., and de Heer, W. A. (1999). Electrostatic deflections and electromechanical resonances of carbon nanotubes, *Science*, **283**, pp. 1513–1516.

12. Gaillard, J., Skove, M. J., Ciocan, R., and Rao, A. M. (2006) Electrical detection of oscillations in microcantilevers and nanocantilevers, *Rev. Sci. Instrum.*, **77**, p. 073907.
13. Keskar, G., Elliott, B., Gaillard, J., Skove, M. J., Rao, A. M. (2008) Using electric actuation and detection of oscillations in microcantilevers for pressure measurements, *Sens. Actuators A*, **147**, pp. 203–209.
14. Salvetat, J.-P., Andrew, G., Briggs, D., Bonard, J.-M., Bacsa, R. R., Kulik, A. J., Stökli, T., Burnham, N. A., and Forró, L. (1999). Elastic and shear moduli of single-walled carbon nanotube ropes. *Phys. Rev. Lett.*, **82**(5), pp. 944–947.
15. Kreyszig, E. (1993). *Advanced Engineering Mathematics (7th edn.)*, (Wiley, New York), pp. 643–644.
16. Treacy, M. M. J., Ebbesen, T. W., and Gibson, J. M. (1996). Exceptionally high Young's modulus observed for individual carbon nanotubes, *Nature*, **381**, pp. 678–680.
17. Krishnan, A., Dujardin, E., Ebbesen, T. W., Yianilos, P. N., and Treacy, M. M. J. (1998). Young's modulus of single-walled nanotubes, *Phys. Rev. B*, **58**, pp. 14013–14019.
18. Chopra, N. G., and Zettl A. (1998). Measurement of the elastic modulus of a multi-wall boron nitride nanotube, *Solid State Commun.*, **105**, pp. 297–300.
19. Wang, Z. L., Poncharal, P., and de Heer, W. A. (2000). Measuring physical and mechanical properties of individual carbon nanotubes by *in situ* TEM, *J. Phys. Chem. Solid.*, **61**, pp. 1025–1030.
20. Griffths, D. J., and Li, Y. (1996). Charge density on a conducting needle, *Am. J. Phys.*, **64**, pp. 706–714.
21. Meirovich, L. (1986). *Elements of Vibration Analysis, 2nd ed.* (McGraw-Hill, New York).
22. Ruoff, R. S., and Lorents, D. C. (1995). Mechanical and thermal properties of carbon nanotubes, *Carbon*, **33**(7), pp. 925–930.
23. Kuzumaki, T., Hayashi, T., Ichinose, H., Miyazawa, K., Ito, K and Ishida, Y. (1998). *In-situ* observed deformation of carbon nanotubes, *Phil. Mag. A*, **77**(6), pp. 1461–1469.
24. Wong, E. W., Sheehan, P. E., and Lieber, C. M. (1997). Nanobeam mechanics: elasticity, strength, and toughness of nanorods and nanotubes, *Science*, **277**(5334), pp. 1971–1975.
25. Sheehan, P. E., and Lieber, C. M. (1996). Nanomachining, Manipulation and Fabrication by Force Microscopy, *Nanotechnology*, **7**, pp. 236–240.

26. Salvetat J.-P., Kulik, A. J., Bonard J.-M., Briggs, G. A. D., Stöckli, T., Méténier, K., Bonnamy, S., Béguin, F., Burnham, N. A., and Forró L., (1999). Elastic Modulus of Ordered and Disordered Multiwalled Carbon Nanotubes *Adv. Mater.*, **11**, pp. 161–165.
27. Kis, A., Csányi G., Salvetat J.-P., Lee, T.-N., Couteau, E., Kulik. A. J., Benoit, W., Brugger, J., and Forró, L. (2004). Reinforcement of single-walled carbon nanotube bundles by intertube bridging. *Nature Mater.*, **3**, pp. 153–157.
28. Kis, A., Jensen, K., Aloni, S., Mickelson, W., and Zettl, A. (2006). Interlayer forces and ultralow sliding friction in multiwalled carbon nanotubes, *Phys. Rev. Lett.*, **97**, pp. 025501_1–025501_4.
29. Yu, M.-F., Lourie, O., Dyer, M. J., Moloni, K., Kelly, T. F., and Ruoff, R. S. (2000). Strength and Breaking Mechanism of Multiwalled Carbon Nanotubes Under Tensile Load. *Science*, **287**(5453), pp. 637–640.
30. Yu, M.-F., Files, B. S., Arepalli, S., and Ruoff, R. S. (2000). Tensile loading of ropes of single wall carbon nanotubes and their mechanical properties, *Phys. Rev. Lett.*, **84**, pp. 5552–5555.
31. Iijima, S., Brabec, C., Maiti, A., and Bernholc, J. (1996). Structural flexibility of carbon nanotubes, *J. Chem. Phys.*, **104**, pp. 2089–2092.
32. Yakobson, B. I., Brabec, C. J., and Bernholc, J. (1996). Nanomechanics of carbon tubes: instabilities beyond linear response. *Phys. Rev. Lett.*, **76**(14), pp. 2511–2514.
33. Hernández, E., Goze, C., Bernier, P., and Rubio, A. (1998). Elastic properties of C and BxCyNz composite nanotubes, *Phys. Rev. Lett.*, **80**, pp. 4502–4505.
34. Srivastava, D., Menon, M., and Cho, K. (1999). Nanoplasticity of single-wall carbon nanotubes under uniaxial compression, *Phys. Rev. Lett.*, **83**, pp. 2973–2976.
35. Sánchez-Portal, D., Artacho, E., Soler, J. M., Rubio, A., and Ordejón, P. (1998). *Ab initio* structural, elastic, and vibrational properties of carbon nanotubes, *Phys. Rev. B*, **59**, pp. 12678–12688.
36. Hohenberg, P., and Kohn, W. (1964). Inhomogeneous electron gas, *Phys. Rev.*, **136**, pp. B864–B871.
37. Kohn, W., and Sham, L. J. (1965). Self-consistent equations including exchange and correlation effects, *Phys. Rev.*, **140**, pp. A1133–A1138.
38. Car, R., and Parrinello, M. (1985). Unified approach for moleculardynamics and density-functional theory, *Phys. Rev. Lett.*, **55**, pp. 2471–2474.

39. Payne, M. C., Teter, M. P., Allan, D. C., Arias. T. A., and Joannopoulos J. D. (1992). Iterative minimization techniques for *ab initio* total-energy calculations: molecular-dynamics and conjugate gradients, *Rev. Mod. Phys.*, **64**, pp. 1045–1097.
40. Slater, J. C., and Koster, G. F. (1954). Simplified LCAO Method for the Periodic Potential Problem, *Phys. Rev.*, **94**, pp. 1498–1594.
41. Verlet, L. (1967). Computer "experiments" on classical fluids. I. Thermodynamical properties of lennard-jones molecules, *Phys. Rev.*, **159**, pp. 98–103.
42. Cao, G., and Chen, X. (2006a). Mechanisms of nanoindentation on single-walled carbon nanotubes: the effect of nanotube length, *J. Mater. Res.*, **21**, pp. 1048–1070, and Cao, G., Chen, X., and Kysar, J. W. (2006a). Numerical analysis of the radial breathing mode of individual armchair and zigzag single-walled carbon nanotubes under deformation, *J. Appl. Phys.*, **100**, pp. 124305–124314.
43. Kundin, K. N., Scuseria, G. E., and Yakobson, B. I. (2001). C_2F, BN, and C nanoshell elasticity from *ab initio* computations, *Phys. Rev. B*, **64**, pp. 235406_1–235406_10.
44. Porezag, D., Frauenheim, Th. Kohler, Th., Seifert, G., and Kascher, R. (1995). Construction of tighet-binding-like potentials on basis of density-functional theory: application to carbon, *Phys. Rev. B*, **51**, pp. 12947–12957.
45. Nardelli, M., and Bernholc, J. (1999). Mechanical deformations and coherent transport in carbon nanotubes, *Phys. Rev. B*, **60**, pp. 16338–16341.
46. Zhang, P., Lammert, P. E., Crespi, V. H. (1998). Plastic deformations of carbon nanotubes, *Phys. Rev. Lett.*, **81**, pp. 5346–5349.
47. Sun, H. (1998). COMPASS: an *ab initio* force field optimized for condensed-phase applications. Overview with details on alkane and benzene compounds, *J. Phys. Chem. B*, **102**, pp. 7338–7364.
48. Cao, G., and Chen, X. (2006). Buckling behaviors of the single-walled carbon nanotubes and a targeted-molecular dynamics simulation, *Phys. Rev. B*, **74**, pp. 165422_1–165422_10.
49. Cao, G., and Chen, X. (2006). The effect of displacement increment on the axial compressive buckling behavior of single-walled carbon nanotubes, *Nanotechnology*, **17**, pp. 3844–3855.
50. Cao, G., and Chen, X. (2007). The effects of chirality and boundary conditions on the mechanical properties of single-walled carbon nanotubes, *Int. J. Solids Struct.*, **44**, pp. 5447–5465.

51. Abell, G. C. (1985). Empirical chemical pseudopotential theory of molecular and metallic bonding, *Phys. Rev. B*, **31**, pp. 6184–6196.

52. Tersoff, J. (1987). New empirical approach for the structure and energy of covalent systems, *Phys. Rev. B*, **37**(12), pp. 6991 7000.

53. Tersoff, J. (1988). Empirical interatomic potential for carbon, with applications to amorphous carbon, *Phys. Rev. Lett.*, **61**, pp. 2879–2882.

54. Tersoff, J. (1988) New empirical approach for the structure and energy of covalent systems, *Phys. Rev. B*, **37**, pp. 6991–7000.

55. Tersoff, J. (1989). Modeling solid-state chemistry: interatomic potentials for multicomponent systems, *Phys. Rev. B*, **39**, pp. 5566–5568.

56. Brenner, D. W. (1990). Empirical potential for hydrocarbons for use in simulating the chemical vapor deposition of diamond films, *Phys. Rev. B*, **42**, pp. 9458–9471.

57. Brenner, D. W., Shenderova, O. A., Harrison, J. A., Stuart, S, J., Ni, B., and Sinnott, S. B. (2002). A second-generation reactive empirical bond order (REBO) potential energy expression for hydrocarbons, *J. Phys. Condens. Mat.*, **14**, pp. 783–802.

58. Srivastava, D., Wei, C., and Cho, K. (2003). Nanomechanics of carbon nanotubes and composites, *Appl. Mech. Rev.*, **56**, pp. 215–230.

59. Robertson, D. H., Brenner, D. W., and Mintmire J. W. (1992). Energetics of nanoscale graphitic tubules, *Phys. Rev. B*, **45**(21), pp. 12592–12595.

60. Tibbetts, G. G. (1983). Why are carbon filaments tubular? *J. Cryst. Growth*, **66**, pp. 632–638.

61. Lu, J. P. (1997). Elastic properties of carbon nanotubes and nanoropes, *Phys. Rev. Lett.*, **79**, pp. 1297–1300.

62. Soler, J. M, Artacho, E., Gale, J. D., García, A., Junquera, J., Ordejón, P., and Sánchez-Portal, D. (2002). The SIESTA method for *ab initio* order-N materials simulation, *J. Phys. Condens. Mat.*, **14**, pp. 2745–2779.

63. Hernández, E., Goze, C., Bernier, P., and Rubio, A. (1999). Elastic properties of single-wall nanotubes, *Appl. Phys. A*, **68**, pp. 287–292.

64. Chandraseker, K., and Mukherjee, S. (2007). Atomistic-continuum and *ab initio* estimation of the elastic moduli of single-walled carbon nanotubes *Comp. Mater. Sci.*, **40**, pp. 147–158.

65. Kelly B. K. (1981). *Physics of Graphite* (Applied Science, London).

66. Dresselhaus, M. S., and Dresselhaus, G. (1998). *Graphite Fibers and Filaments* (Springer, Berlin, Heidelberg).

67. Iijima, S., Brabec, C., Maiti, A., and Bernholc, J. (1995). Structural flexibility of carbon nanotubes, *J. Chem. Phys.*, **104**, pp. 2082–2093.

68. Cornwell, C. F., and Wille, L. T. (1997). Elastic properties of single-walled carbon nanotubes in compression, *Solid State Commun.*, **101**, pp. 555–558.

69. Popov, V. N., Van Doren V. E., and Balkanski, M. (2000). Elastic properties of single-walled carbon nanotubes, *Phys. Rev. B*, **61**, pp. 3078–3084.

70. Chang, T., and Gao, H. (2003). Size-dependent elastic properties of a single-walled carbon nanotube via a molecular mechanics model, *J. Mech. Phys. Solids*, **51**, pp. 1059–1074.

71. Arroyo, M., and Belytschko, T. (2005). Continuum mechanics modeling and simulation of carbon nanotubes, *Mechanica*, **40**, pp. 455–469.

72. Landau, L. D., and Lifshitz, E. M. (1986). *Theory of Elasticity, 3rd ed.* (Pergamon, New York).

73. Timoshenko, S., and Goodier, S. N. (1970). *Theory of Elasticity, 3rd ed.* (McGraw-Hill).

74. Odegard, G. M., Gates, T. S., Nicholson, L. M., Wise, K. E. (2002). Equivalent-continuum modeling of nano-structured materials, *Comp. Sci. Tech.*, **62**, pp. 1869–1880.

75. Jin, Y., and Yuan, F. G. (2003). Simulation of elastic properties of single-walled carbon nanotubes, *Comp. Sci. Tech.*, **63**, pp. 1507–1515.

76. Vodenitcharova, T., and Zhang, L. C. (2003). Effective wall thickness of a single-walled carbon nanotube, *Phys. Rev. B*, **68**, pp. 165401_1–165401_4.

77. Goupalov, S. V., (2005). Continuum model for long-wavelength phonons in two-dimensional graphite and carbon nanotubes, *Phys. Rev. B*, **71**, pp. 085420_1–085420_7.

78. Wang, L., Zheng, Q., Liu, J. Z., and Jiang, Q., (2005). Size Dependence of the thin-shell model for carbon nanotubes, *Phys. Rev. Lett.*, **95**, pp. 105501_1–105501_4.

79. Harik, V. M. (2002). Mechanics of carbon nanotubes: applicability of the continuum-beam models, *Comput. Mater. Sci.*, **24**, pp. 328–342.

80. Arroyo, M., and Belytschko, T. (2002). An atomistic-based finite deformation membrane for single layer crystalline films, *J. Mech. Phys. Solids*, **50**, pp. 1941–1977.

81. Zhang, P., Huang, Y., Geubelle, P. H., Klein, P. A., and Hwang, K. C. (2002). The elastic modulus of single-wall carbon nanotubes: a continuum analysis incorporating interatomic potentials, *Int. J. Solids Struct.*, **39**, pp. 3893–3906.

82. Li, C., and Chou, T.-W. (2003). A structural mechanics approach for the analysis of carbon nanotubes, *Int. J. Solids Struct.*, **40**, pp. 2487–2499.
83. Lier, G. V., Alsenoy, C. V., Doren, V. V., and Geerlings, P. (2000). *Ab initio* study of the elastic properties of single-walled carbon nanotubes and graphene, *Chem. Phys. Lett.*, **326**, pp. 181–185.
84. Ruoff, R. S., and Lorents, D. C. (1995). Mechanical and thermal properties of carbon nanotubes, *Carbon*, **33**, pp. 925–930.
85. Lu, J. P. (1997). Elastic properties of single and multilayered nanotubes, *J. Phys. Chem. Solid.*, **58**, pp. 1649–1652.
86. Timoshenko, S., and Woinowsky-Krieger, W. (1959). *Theory of Plates and Shells, 2nd ed.* (McGraw-Hill).
87. Tserpes, K. J., and Papanikos, P. (2005). Finite element modeling of single-walled carbon nanotubes, *Composites B*, **36**, pp. 468–477.
88. See the official Web site: http://www.ansys.com/.
89. Kalamkarov, A. L., Georgiades, A. V., Rokkam, S. K., Veedu, V. P., Ghasemi-Nejhad, M. N. (2006). Analytical and numerical techniques to predict carbon nanotubes properties, *Int. J. Solids Struct.*, **43**, pp. 6832–6854.
90. Wan, H., and Delale, F. (2010). A structural mechanics approach for predicting the mechanical properties of carbon nanotubes, *Meccanica*, **45**, pp. 43–51.
91. Ru, C. Q. (2000) Column buckling of multiwalled carbon nanotubes with interlayer radial displacements, *Phys. Rev. B*, **62**, pp. 16962–16967.
92. Lucas, A. A., Lambin, P. H., and Smalley, R. E. (1993). On the energetics of tubular fullerenes, *J. Phys. Chem. Solid.*, **54**, pp. 587–593.
93. Tu, Z. C., and Yang, Z. C. O. (2002). Single-walled and multiwalled carbon nanotubes viewed as elastic tubes with the effective Young's moduli dependent on layer number, *Phys. Rev. B*, **65**, pp. 233407_1–233407_4.
94. Huang, Y., Wu, J., and Hwang, K. (2006). Thickness of graphene and single-wall carbon nanotubes, *Phys. Rev. B*, **74**, pp. 245413_1–245413_9.
95. Chen, X., and Cao, G. (2006). A structural mechanics approach of single-walled carbon nanotubes generalized from atomistic simulation, *Nanotechnology*, **17**, pp. 1004–1015.
96. Pantano, A., Boyce, M. C., and Parks, D. M. (2003). Nonlinear structural mechanics based modeling of carbon nanotube deformation, *Phys. Rev. Lett.*, **91**, pp. 145501–145504.

97. Zhou, X., Zhou, J. J., and Ou-Yang, Z. C. (2000). Strain energy and Young's modulus of single-wall carbon nanotubes calculated from electronic energy-band theory, *Phys. Rev. B*, **62**, pp. 13692–13696.

98. Lu, Q., Arroyo, M., and Huang, R. (2009). Elastic bending modulus of monolayer graphene, *J. Phys. D: Appl. Phys.*, **42**, 102002.

99. Jiang, H., Zhang, P., Liu, B., Huang, Y., Geubelle, P. H., Gao, H., and Hwang, K. C. (2003). The effect of nanotube radius on the constitutive model for carbon nanotubes, *Comput. Mater. Sci.*, **28**, pp. 429–442.

100. Jiang, H., Feng, X. Q., Huang, Y., Hwang, K. C., and Wu, P. (2004a). Defect nucleation in carbon nanotubes under tension and torsion: Stone–Wales transformation, *Comput. Methods Appl. Mech. Eng.*, **193**, pp. 3419–3429.

101. Zhang, P., Huang, Y., Gao, H., and Hwang, K. C. (2002a). Fracture nucleation in single-wall carbon nanotubes under tension: a continuum analysis incorporating interatomic potentials, *J. Appl. Mech. Rev.*, **69**, pp. 454–458.

102. Zhang, P., Huang, Y., Geubelle, P. H., and Hwang, K. C. (2002b). On the continuum modeling of carbon nanotubes, *Acta Mech. Sin.*, **18**, pp. 528–536.

103. Jiang, H., Huang, Y., Zhang, P., and Hwang, K. C. (2006). *Fracture Nucleation in Single-Wall Carbon Nanotubes: The Effect of Nanotube Charality* (Springer, Dordrecht).

104. Zhang, P., Huang, Y., Geubelle, P. H., Klein, P. A., and Hwang, K. C. (2002c). The elastic modulus of single-wall carbon nanotubes: a continuum analysis incorporating interatomic potentials, *Int. J. Solids Struct.*, **39**, pp. 3893–3906.

105. Zhang, P., Jiang, H., Huang, Y., Geubelle, P. H., and Hwang, K. C. (2004). An atomistic-based continuum theory for carbon nanotubes: analysis of fracture nucleation, *J. Mech. Phys. Solids*, **52**, pp. 977–998.

106. Shi, D. L., Feng, X. Q., Jiang, H., Huang, Y., and Hwang, K. C. (2005). Multiscale analysis of fracture of carbon nanotubes embedded in composites, *Int. J. Fract.*, **134**, pp. 369–386.

107. Song, J., Huang, Y., Jiang, H., Hwang, K. C., and Yu, M. F. (2006a). Deformation and bifurcation analysis of boron-nitride nanotubes, *Int. J. Mech. Sci.*, **48**, pp. 1197–1207.

108. Song, J., Jiang, H., Shi, D. L., Feng, X. Q., Huang, Y., Yu, M. F., and Hwang, K. C. (2006b). Stone–Wales transformation: precurson of fracture in carbon nanotubes, *Int. J. Mech. Sci.*, **48**, pp. 1464–1470.

109. Jiang, H., Liu, B., Huang, Y., and Hwang, K. C. (2004b). Thermal expansion of single-wall carbon nanotubes, *J. Eng. Mater. Technol.*, **126**, pp. 265–270.

110. Jiang, H., Huang, Y., and Hwang, K. C. (2005). A finite-temperature continuum theory based on the interatomic potential, *J. Eng. Mater. Technol.*, **127**, pp. 408–416.

111. Johnson, H. T., Liu, B., and Huang, Y. (2004). Electron transport in deformed carbon nanotubes, *J. Eng. Mater. Technol.*, **126**, pp. 222–229.

112. Liu, B., Huang, Y., Jiang, H., Qu, S., and Hwang, K. C. (2004a). The atomic-scale finite-element method, *Comput. Methods Appl. Mech. Eng.*, **193**, pp. 1849–1864.

113. Chandraseker, K., Mukherjee, S., and Mukherjee, Y. X. (2006). Modifications to the Cauchy–Born rule: applications in the deformation of single-walled carbon nanotubes, *Int. J. Solids Struct.*, **43**, pp. 7128–7144.

114. Liu, B., Jiang, H., Johnson, H. T., and Huang, Y. (2004b). The influence of mechanical deformation on the electrical properties of single-wall carbon nanotubes, *J. Mech. Phys. Solids*, **52**, pp. 1–26.

115. Arroyo, M., and Belytschko, T. (2004). Finite crystal elasticity of carbon nanotubes based on the exponential Cauchy–Born rule, *Phys. Rev. B*, **69**, pp. 115415_1–115415_11.

116. Srolovitz, D. J., Safran, S. A., and Tenne, R. (1994). Elastic equilibrium of curved thin films, *Phys. Rev. E*, **49**, pp. 5260–5270.

117. Cao, G., and Chen, X. (2006c). Buckling of single-walled carbon nanotubes under bending: molecular dynamics and finite element simulations, *Phys. Rev. B*, **73**, pp. 155435_1–155435_10.

118. Pantano, A., Parks, D. M., and Boyce, M. C. (2004). Mechanics of deformation of single- and multiwall carbon nanotubes, *J. Mech. Phys. Solids*, **52**, pp. 789–821.

119. Popov, V. N., Van Doren, V. E., and Balkanski, M. (2000). Elastic properties of crystals of single-walled carbon nanotubes, *Solid State Commun.*, **114**, pp. 395–399.

120. Saether, E., Frankland, S. J. V., and Pipes, R. B. (2003). Transverse mechanical properties of single-walled carbon nanotube crystals. Part I: determination of elastic moduli, *Compos. Sci. Technol.*, **63**, pp. 1543–1550.

121. Wang, C. M., Zhang, Y. Y., Xiang, Y., and Reddy, J. N. (2010). Recent studies on buckling of carbon nanotubes, *Appl. Mech. Rev.*, **63**, 030804_1–030804_18.

122. Shima, H. (2012). Buckling of carbon nanotubes: A state of the art review, *Materials*, **5**, pp. 47–84.
123. Brush, D. O., and Almroth, B. O. (1975). *Buckling of bars, plates, and shells* (McGraw-Hill)
124. Ru, C. Q. (2000). Effect of Van der Waals forces on axial buckling of a doublewalled carbon nanotube, *J. Appl. Phys.*, **87**, pp. 7227–7231.
125. Ru, C. Q. (2001). Axially compressed buckling of a doublewalled carbon nanotube embedded in an elastic medium, *J. Mech. Phys. Solids*, **49**, pp. 1265–1279.
126. Ru, C. Q. (2001). Degraded axial buckling strain of multiwalled carbon Nanotubes due to interlayer slips, *J. Appl. Phys.*, **89**, pp. 3426–3433.
127. Ni, B., Sinnott, S. B., Mikulski, P. T., and Harrison, J. A. (2002). Compression of carbon nanotubes filled with C_{60}, CH_4, or Ne: predictions from molecular dynamics simulations, *Phys. Rev. Lett.*, **88**(20), pp. 205505_1–205505_4.
128. Buehler, M. J., Kong, J., and Gao, H. J. (2004). Deformation mechanism of very long single-wall carbon nanotubes subject to compressive loading, *ASME J. Eng. Mater. Technol.*, **126**, pp. 245–249.
129. Waters, J. F., Riester, L., Jouzi, M., Guduru, P. R., and Xu, J. M. (2004). Buckling instabilities in multiwalled carbon nanotubes under uniaxial compression, *Appl. Phys. Lett.*, **85**, pp. 1787–1789.
130. Pantano, A., Boyce, M. C., and Parks, D. M. (2004). Mechanics of axial compression of single- and multi-wall carbon nanotubes, *ASME J. Eng. Mater. Technol.*, **126**, pp. 279–284.
131. Waters, J. F., Guduru, P. R., Jouzi, M., Xu, J. M., Hanlon, T., and Suresh, S. (2005). Shell buckling of individual multi-walled carbon nanotubes using nanoindentation, *Appl. Phys. Lett.*, **87**(10), pp. 103109_1–103109_3.
132. Sears, A., and Batra, R. C. (2006), Buckling of multiwalled carbon nanotubes under axial compression, *Phys. Rev. B*, **73**, pp. 085410_1–085410_11.
133. Zhang, Y. Y., Wang, M., and Tan, V. B. C. (2008). Examining the effects of wall numbers on buckling behavior and mechanical properties of multiwalled carbon nanotubes via molecular dynamics simulations, *J. Appl. Phys.*, **103**, pp. 053505_1–053505_9.
134. Falvo, M. R., Clary G. J., Taylor II, R. M., Chi V., Brooks Jr, F. P., Washburn S., and Superfine, R. (1997). Bending and buckling of carbon nanotubes under large strain, *Nature*, **389**, pp. 582–584.

135. Duan, X. J., Tang, C., Zhang, J., Guo, W. L., and Liu, Z. F. (2007). Two distinct buckling modes in carbon nanotube bending, *Nano Lett.*, **7**, pp. 143–148.

136. Shibutani, Y., and Ogata, S. (2004). Mechanical integrity of carbon nanotubes for bending and torsion, *Modell. Simul. Mater. Sci. Eng.*, **12**, pp. 599–610.

137. Kutana, A., and Giapis, K. P. (2006). Transient deformation regime in bending of single-walled carbon nanotubes, *Phys. Rev. Lett.*, **97**, pp. 245501_1–245501_4.

138. Yang, H. K., and Wang, X. (2006). Bending stability of multi-wall carbon nanotubes embedded in an elastic medium, *Modell. Simul. Mater. Sci. Eng.*, **14**, pp. 99–116.

139. Wang, X., and Yang, H. K. (2006). Bending stability of multiwalled carbon nanotubes, *Phys. Rev. B*, **73**(8), pp. 085409_1–085409_8.

140. Wang, Q., Liew, K. M., He, X. Q., and Xiang, Y. (2007). Local buckling of carbon nanotubes under bending, *Appl. Phys. Lett.*, **73**(91), pp. 093128_1–093128_3.

141. Jeong, B. W., Lim, J. K., and Sinnott, S. B. (2007). Turning stiffness of carbon nanotube systems, *Appl. Phys. Lett.*, **91**, pp. 093102_1–093102_3.

142. Yang, H. K., and Wang, X. (2007). Torsional buckling of multi-wall carbon nanotubes embedded in an elastic medium, *Compos. Struct.*, **77**, pp. 182–192.

143. Zhang, Y. Y., and Wang, C. M. (2008). Torsional responses of double-walled carbon nanotubes via molecular dynamics simulations, *J. Phys.: Condens. Matter*, **20**(45), pp. 455214_1–455214_7.

144. Wang, Q. (2008). Torsional buckling of double-walled carbon nanotubes, *Carbon*, **46**, pp. 1172–1174.

145. Wang, Q. (2009). Transportation of hydrogen molecules using carbon nanotube in torsion, *Carbon*, **47**, pp. 1870–1873.

146. Wang, X., Sun, B., and Yang, H. K. (2006). Stability of multi-walled carbon nanotubes under combined bending and axial compression loading, *Nanotechnology*, **17**, pp. 815–823.

147. Lu, Y. J., and Wang, X., (2006). Combined torsional buckling of multi-walled carbon nanotubes, *J. Phys. D*, **39**, pp. 3380–3387.

148. Wang, X., Lu, G. X., and Lu, Y. J., (2007), Buckling of embedded multi-walled carbon nanotubes under combined torsion and axial loading, *Int. J. Solids Struct.*, **44**, pp. 336–351.

149. Zhang, C. L., and Shen, H. S. (2007). Buckling and postbuckling of single-walled carbon nanotubes under combined axial compression and torsion in thermal environments, *Phys. Rev. B*, **75**, pp. 045408_1–045408_7.

150. Jeong, B. W., Lim, J. K., and Sinnott, S. B. (2008). Tuning the torsional properties of carbon nanotube systems with axial prestress, *Appl. Phys. Lett.*, **92**, pp. 253114_1–253114_3.

151. Bower, C., Rosen, R., Jin, L., Han, J., and Zhou, O. (1999). Deformation of carbon nanotubes in nanotube–polymer composites, *Appl. Phys. Lett.*, **74**, pp. 3317–3319.

152. Arroyo, M., and Belytschko, T. (2003). Nonlinear mechanical response and rippling of thick multiwalled carbon nanotubes, *Phys. Rev. Lett.*, **91**, pp. 215505_1–215505_4.

153. Chang, T., and Hou, J. (2006). Molecular dynamics simulations on buckling of multiwalled carbon nanotubes under bending, *J. Appl. Phys.*, **100**, pp. 114327_1–114327_5.

154. Li, X. Y., Yang, W., and Liu, B. (2007). Bending induced rippling and twisting of multiwalled carbon nanotubes, *Phys. Rev. Lett.*, **98**, pp. 205502_1–205502_4.

155. Arias, I., and Arroyo, M. (2008). Size-dependent nonlinear elastic scaling of multiwalled carbon nanotubes, *Phys. Rev. Lett.*, **100**, pp. 085503_1–085503_4.

156. Arroyo, M., and Arias, I. (2008). Rippling and a phase-transforming mesoscopic model for multiwalled carbon nanotubes, *J. Mech. Phys. Solids*, **56**, pp. 1224–1244.

157. Huang, X., Zou, J., and Zhang, S. L. (2008). Bilinear responses and rippling morphologies of multi-walled carbon nanotubes under torsion, *Appl. Phys. Lett.*, **93**, pp. 031915_1–031915_3.

158. Zou, J., Huang, X., Arroyo, M., and Zhang, S. L. (2009). Effective coarse-grained simulations of super-thick multi-walled carbon nanotubes under torsion, *J. Appl. Phys.*, **105**, pp. 033516_1–033516_8.

159. Huang, X., Yuan, H. Y., Hsia, K. J., and Zhang, S. L. (2010). Coordinated buckling of thick multi-walled carbon nanotubes under uniaxial compression, *Nano Res.*, **3**(1), pp. 32–42.

160. Knechtel, W. H., Dusberg, G. S., Blau, W. J., Hernandez, E., and Rubio, A. (1998). Reversible bending of carbon nanotubes using a transmission electron microscope, *Appl. Phys. Lett.*, **73**, pp. 1961–1963.

161. Lourie, O., Cox, D. M., and Wagner, H. D. (1998). Buckling and collapse of embedded carbon nanotubes, *Phys. Rev. Lett.*, **81**, pp. 1638–1641.

162. Thomas, W., Tombler, Zhou, C., Alexseyev, L., Kong, J., Dai, H., Liu, L., Jayanthi, C. S., Tang, M., and Wu, S. (2000), Reversible electromechanical characteristics of carbon nanotubes under local-probe manipulation, *Nature*, **405**, pp. 769–772.

163. Lee, C., Wei, X. D., Kysar, J. W., and Hone, J. (2008). Measurement of the elastic properties and intrinsic strength of monolayer graphene, *Science*, **321**, pp. 385–388.

164. Li, X., Maute, K., Dunn, M. L., and Yang, R. (2010), Strain effects on the thermal conductivity of nanostructures, *Phys. Rev. B*, **81**, pp. 245318_1–245318_11.

165. Yang, W., Wang, R. Z., and Yan, H. (2008). Strain-induced Raman-mode shift in single-wall carbon nanotubes: calculation of force constants from molecular-dynamics simulations, *Phys. Rev. B*, **77**, pp. 195440_1–195440_5.

166. Tang, C., Guo, W., and Chen, C. (2010). Bending manipulation induced sp^2-sp^3 bond transition in carbon nanotubes, *J. Appl. Phys.*, **108**, pp. 026108_1–026108_3.

167. Rochefort, A., Avouris, P., Lesage, F., and Salahub D. R. (1999). Electrical and mechanical properties of distorted carbon nanotubes, *Phys. Rev. B*, **60**, pp. 13824–13830.

168. Bozovic, D., Bockrath, M., Hafner, J. H., Lieber, C. M., Park, H., and Tinkham, M. (2001). Electronic properties of mechanically induced kinks in single-walled carbon nanotubes, *Appl. Phys. Lett.*, **78**(23), pp. 3693–3695.

169. Farajian, A. A., Yakobson, B. I., Mizuseki, H., and Kawazoe, Y. (2003). Electronic transport through bent carbon nanotubes: nanoelectromechanical sensors and switches, *Phys. Rev. B*, **67**, pp. 205423_1–205423_5.

170. Peters, M. J., McNeil, L. E., Lu, J. P., and Kahan, D. (2000). Structural phase transition in carbon nanotube bundles under pressure, *Phys. Rev. B*, **61**, pp. 5939–5944.

171. Liu, J. Z., Zheng, Q., and Jiang, Q. (2001). Effect of a rippling mode on resonances of carbon nanotubes, *Phys. Rev. Lett.*, **86**(21), pp. 4843–4846.

172. Chang, T. (2007). Torsional behavior of chiral single-walled carbon nanotubes is loading direction dependent, *Appl. Phys. Lett.*, **90**, pp. 201910_1–201910_3.

173. Geng, J., and Chang, T. (2006). Nonlinear stick-spiral model for predicting mechanical behavior of single-walled carbon nanotubes, *Phys. Rev. B*, **74**, pp. 245428_1–245428_13.

174. Liang, H., and Upmanyu, M. (2006). Axial-strain-induced torsion in single-walled carbon nanotubes, *Phys. Rev. Lett.*, **96**, pp. 165501_1–165501_4.

175. Cohen-Karni, T., Segev, L., Srur-Lavi, O., Cohen, S. R., and Joselevich, E. (2006). Torsional electromechanical quantum oscillations in carbon nanotubes, *Nat. Nanotechnol.*, **1**, pp. 36–41.

176. Hall, A. R., Falvo, M. R., Superfine, R., and Washburn, S. (2007). Electromechanical response of single-walled carbon nanotubes to torsional strain in a self-contained device, *Nat. Nanotech.*, **2**, pp. 413–416.

177. Nagapriya, K. S., Berber, S., Cohen-Karni, T., Segev L., Srur-Lavi, O., Tománek, D., and Joselevich, E., (2008). Origin of torsion-induced conductance oscillations in carbon nanotubes, *Phys. Rev. B*, **78**, pp. 165417_1–165417_5.

178. Hall, A. R., Falvo, M. R., Superfine, R., and Washburn, S. (2008). A self-sensing nanomechanical resonator built on a single-walled carbon nanotube, *Nano Lett.*, **8**, pp. 3746–3749.

179. Palaci, I., Fedrigo, S., Brune, H., Klinke, C., Chen, M., and Riedo, E., (2005). Radial elasticity of multiwalled carbon nanotubes, *Phys. Rev. Lett.*, **94**, pp. 175502_1–175502_4.

180. J. Tang, J. C. Qin, T. Sasaki, M. Yudasaka, A. Matsushita, and S. Iijima, (2000). Compressibility and polygonization of single-walled carbon nanotubes under hydrostatic pressure, *Phys. Rev. Lett.*, **85**, pp. 1887–1889.

181. Peters, M. J., McNeil, L. E., Lu J. P., and Kahn, D. (2000). Structural phase transition in carbon nanotube bundles under pressure, *Phys. Rev. B*, **61**(9), pp. 5939–5944.

182. Sharma, S. M., Karmakar, S., Sikka, S. K., Teredesai, P. V., Sood, A. K., Govindaraj, A., and Rao, C. N. R. (2001). Pressure-induced phase transformation and structural resilience of single-wall carbon nanotube bundles, *Phys. Rev. B*, **63**, pp. 205417_1–205417_5.

183. Rols, S., Gontcharenko, I. N., Almairac, R., Sauvajol, J. L., and Mirebeau, I. (2001). Polygonization of single-wall carbon nanotube bundles under high pressure, *Phys. Rev. B*, **64**, pp. 153401_1–153401_4.

184. Reich, S., Thomsen, C., and Ordejón P. (2003). Elastic properties of carbon nanotubes under hydrostatic pressure, *Phys. Rev. B*, **65**, 153407_1–153407_4.

185. Pantano, A., Parks, D. M., and Boyce, M. C. (2004). Mechanics of deformation of single- and multi-wall carbon nanotubes, *J. Mech. Phys. Solid.*, **52**(4), pp. 789–821.

186. Elliott, J. A., Sandler, L. K. W., Windle, A. H., Young, R. J., and Shaffer, M. S. P. (2004). Collapse of single-wall carbon nanotubes is diameter dependent, *Phys. Rev. Lett.*, **92**, pp. 095501_1–095501_4.

187. Tangney, P., Capaz, R. B., Spataru, C. D., Cohen, M. L., and Louie, S. G. (2005). Structural transformations of carbon nanotubes under hydrostatic pressure, *Nano Lett.*, **5**(11), pp. 2268–2273.

188. Gadagkar, V., Maiti, P. K., Lansac, Y., Jagota, A., and Sood, A. K. (2006). Collapse of double-walled carbon nanotube bundles under hydrostatic pressure, *Phys. Rev. B*, **73**, pp. 085402_1–085402_6.

189. Wang, X., and Yang, H. K. (2006). Bending stability of multiwalled carbon nanotubes, *Phys. Rev. B*, **73**(8), pp. 085409_1–085409_8.

190. Zhang, S., Khare, R., Belytschko, T., Hsia, K. J., Mielke, S. L., and Schatz, G. C. (2006). Transition states and minimum energy pathways for the collapse of carbon nanotubes, *Phys. Rev. B*, **73**(7), pp. 075423_1–075423_7.

191. Hasegawa, M., and Nishidate, K. (2006). Radial deformation and stability of single-wall carbon nanotubes under hydrostatic pressure, *Phys. Rev. B*, **74**, pp. 115401_1–115401_11.

192. Yang, X., Wu, G., and Dong, J. (2006). Structural transformations of double-walled carbon nanotube bundle under hydrostatic pressure, *Appl. Phys. Lett.*, **89**(11), pp. 113101_1–113101_3.

193. Christofilos, D., Arvanitidis, J., Kourouklis, G. A., Ves, S., Takenobu, T., Iwasa, Y., and Kataura, H. (2007). Identification of inner and outer shells of double-wall carbon nanotubes using high-pressure Raman spectroscopy, *Phys. Rev. B*, **76**(11), pp. 113402_1–113402_4.

194. Peng, J., Wu, J., Hwang, K. C., Song, J., and Huang, Y. (2008). Can a single-wall carbon nanotube be modeled as a thin shell?, *J. Mech. Phys. Solid.*, **56**(6), pp. 2213–2224.

195. Giusca, C. E., Tison, Y., and Silva, S. R, P. (2008). Evidence for metal-semiconductor transitions in twisted and collapsed double-walled carbon nanotubes by scanning tunneling microscopy, *Nano Lett.*, **8**(10), pp. 3350–3356.

196. Wu, Y., Huang, M., Wang, F., Huang, X. M. H., Rosenblatt, S., Huang, L., Yan, H., O'Brien, S. P., Hone, J., and Heinz, T. F. (2008). Determination of the Young's modulus of structurally defined carbon nanotubes, *Nano Lett.*, **8**(12), pp. 4158–4161.

197. Jeong, B. W., Lim, J. K., and Sinnott, S. B. (2008). Tuning the torsional properties of carbon nanotube systems with axial prestress, *Appl. Phys. Lett.*, **92**(25), pp. 253114_1–_3.

198. Kuang, Y. D., He, X. Q., Chen, C. Y., and Li, G. Q. (2009). Buckling of functionalized single-walled nanotubes under axial compression, *Carbon*, **47**(1), pp. 279–285.

199. Lu, W. B., Liu, B., Wu, J., Xiao, J., Hwang, K. C., Fu, S. Y., and Huang, Y. (2009). Continuum modeling of van der Waals interactions between carbon nanotube walls, *Appl. Phys. Lett.*, **94**(10), pp. 101917_1–101917_3.

200. Venkateswaran, U. D., Rao, A. M., Richter, E., Menon, M., Rinzler, A., Smalley, R. E., and Eklund, P. C. (1999). Probing the single-wall carbon nanotube bundle: Raman scattering under high pressure, *Phys. Rev. B*, **59**(16), pp. 10928–10934.

201. Shima, H., and Sato, M. (2008). Multiple radial corrugations in multiwall carbon nanotubes under pressure, *Nanotechnology*, **19**, pp. 495705_1–495705_8.

202. Zhang, D. B., and Dumitrică, T. (2010). Effective shear-strain driven electromechanical response in helical rippled carbon nanotubes, *Phys. Rev. B*, **82**, pp. 193401_1–193401_4.

203. Charlier, J.-C., Ebbesen, T. W., Lambin, Ph. (1996). Structural and electronic properties of pentagon-heptagon pair defects in carbon nanotubes, *Phys. Rev. B*, **53**, pp. 11108–11113 (1996).

204. Charlier, J.-C. (2002). Defects in carbon nanotubes, *Acc. Chem. Res.*, **35**, pp. 1063–1069.

205. Stone, A. J., and Wales, D. J. (1986). Theoretical studies of icosahedral C60 and some related species, *Chem. Phys. Lett.*, **128**(5–6), pp. 501–503.

206. Nardelli, M. B., Yakobson, B. I., and Bernholc, J. (1998). Mechanism of strain release in carbon nanotubes, *Phys. Rev. B*, **57**, pp. R4277–R4280.

207. Jensen, P., Gale, J., and Blase, X. (2002). Catalysis of nanotube plasticity under tensile strain, *Phys. Rev. B*, **66**, pp. 193403_1–193403_4.

208. Zhao, Q., Nardelli M. B., and Bernholc J. (2002). Ultimate strength of carbon nanotubes: a theoretical study, *Phys. Rev. B*, **65**, pp. 144105_1–144105_6.

209. Ewels, C. P., Heggie, M. I., and Briddon, P. R. (2002). Adatoms and nanoengineering of carbon, *Chem. Phys. Lett.*, **351**, pp. 178–182.

210. Yakobson, B. I. (1998). Mechanical relaxation and "intramolecular plasticity" in carbon nanotubes, *Appl. Phys. Lett.*, **72**, pp. 918–920.

211. Nardelli, M. B., Yakobson B. I., and Bernholc J. (1998). Brittle and ductile behavior in carbon nanotubes, *Phys. Rev. Lett.*, **81**, pp. 4656–4659.

212. Nakayama, Y., Nagataki, A., Suekane, O., Cai, X., and Akita, S. (2005). Current-induced plastic deformation of double-walled carbon nanotubes, *Jpn. J. Appl. Phys.*, **44**, pp. L720–L722.

213. Mori, H., Ogata, S., Li, J., Akita, S., and Nakayama, Y. (2006). Energetics of plastic bending of carbon nanotubes, *Phys. Rev. B*, **74**, pp. 165418_1–165418_5.

214. Zhou, L. G., and Shi, S. Q. (2003). Formation energy of Stone–Wales defects in carbon nanotubes, *Appl. Phys. Lett.*, **83**(7), pp. 1222–1224.

215. Li, Z., Dharap, P., Sharma, P., Nagarajaiah, S., and Yakobson, B. I. (2005). Continuum field model of defect formation in carbon nanotubes, *J. Appl. Phys.*, **97**, pp. 074303_1–074303_8.

216. Walters, D. A., Ericson, L. M., Casavant, M. J., Liu, J., Colbert, D. T., Smith, K. A., and Smalley, R. E. (1999). Elastic strain of freely suspended single-wall carbon nanotube ropes, *Appl. Phys. Lett.*, **74**, pp. 3803–3805.

217. Samsonidze, G. G. (2002). Kinetic theory of symmetry-dependent strength in carbon nanotubes, *Phys. Rev. Lett.*, **88**, pp. 065501_1–065501_4.

218. Ouyang, M., Huang, J.-L., Cheung, C. L., and Lieber, C. M. (2001). Atomically resolved single-walled carbon nanotube intramolecular junctions, *Science*, **291**, pp. 97–100.

219. Hashimoto, A., Suenaga, K., Gloter, A., Urita, K., and Iijima, S. (2004). Direct evidence for atomic defects in graphene layers. *Nature*, **430**, pp. 870–873.

220. Ishigami, M., Choi, H. J., Aloni, S., Louie, S. G., Cohen, M. L., and Zettl, A., (2004). Identifying defects in nanoscale materials, *Phys. Rev. Lett.*, **93**, pp. 196803_1–196803_4.

221. Yoon, M., Han, S., Kim, G., Lee, S. B., Berber, S., Osawa, E., Ihm, J., Terrones, M., Banhart, F., Charlier, J.-C., Grobert, N., Terrones, H., Ajayan, P. M., and Tománek, D. (2004). Zipper mechanism of nanotube fusion: theory and experiment, *Phys. Rev. Lett.*, **92**(7), pp. 075504_1–075504_4.

222. Suenaga, K., Wakabayashi H., Koshino, M., Sato Y., Urita K., and Iijima S. (2007). Imaging active topological defects in carbon nanotubes, *Nat. Nanotechnol.*, **2**, pp. 358–360.

223. Yudasaka, M., Kataura, H., Ichihashi, T., Qin, L.-C., Kar, S., and Iijima, S. (2001). Diameter enlargement of HiPco single-wall carbon nanotubes by heat treatment, *Nano. Lett.*, **1**, pp. 487–489.

224. Meyer, J. C., Kisielowski, C., Erni, R., Rossell, M. D., Crommie, M. F., and Zettl, A. (2008). Direct imaging of lattice atoms and topological defects in graphene membranes, *Nano. Lett.*, **8**, pp. 3582–3586.

225. Danlap, B. I. (1992). Connecting carbon tubules, *Phys. Rev. B*, **46**, pp. 1933–1936.

226. Baldus, M., Tomaslli, M., Meier B. H., and Ernst, R. R. (1994). Broadband polarization-transfer experiments for rotating solids, *Chem. Phys. Lett.*, **223**, pp. 329–336.

227. Motojima, S., Kawaguchi, M., Nozaki, K., and Iwanaga, H. (1990). Growth of regularly coiled filaments by Ni catalyzed pyrolysis of acetylene, and their morphology and extension characteristics, *Appl. Phys. Lett.*, **56**, pp. 321–323.

228. Bervaerts, D., Bzhang, X., Zhang, X. F., Amelinckx, S., Tendeloo, G. V., Landuyt, J. V., Ivanov, V., and Nagy, J. B. (1995). Electron microscopy study of coiled carbon tubules, *Phil. Mag. A*, **71**, pp. 605–630.

229. Hernadi, K., Thiên-Nga, L., and L. Forró. (2001). Growth and microstructure of catalytically produced coiled carbon nanotubes, *J. Phys. Chem. B*, **105**, pp. 12464–12468.

230. Saito, Y., and Yoshikawa, T. (1993). Bamboo-shaped carbon tube filled partially with nickel, *J. Cryst. Growth*, **134**, pp. 154–156.

231. Kovalevski, V. V., and Safronov, A. N. (1998). Pyrolysis of hollow carbons on melted catalyst, *Carbon*, **36**, pp. 963–968.

232. Kiselev, N. A., Sloan, J., Zakharov, D. N., Kukovitskii, E. F., Hutchison, J. L., Hammers, J., and Kotosonov, A. S. (1998). Carbon nanotubes from polyethylene precursors: structure and structural changed caused by thermal and chemical treatment revealed by hrem, *Carbon*, **36**, pp. 1149–1157.

233. Cortijo, A., and Vozmediano, M. A. H. (2007). Effects of topological defects and local curvature on the electronic properties of planar graphene, *Nucl. Phys. B*, **763**, pp. 293–308.

234. Diudea, M. V. (2005). Corannulene and corazulene tiling of nanostructures, *Phys. Chem. Chem. Phys.*, **7**, pp. 3626–3633.

235. Rode, A. V., Gamaly, E. G., and B. Luther-Davies, (2000). Formation of cluster-assembled carbon nano-foam by high-repetition-rate laser ablation, *Appl. Phys. A*, **70**, pp. 135–144.

236. Okada, S., Nakada, K. Kuwabara, K., Daigoku, K., and Kawai, T. (2006). Ferromagnetic spin ordering on carbon nanotubes with topological line defects, *Phys. Rev. B*, **74**, pp. 121412_1–121412_4.

237. Ding, W., Calabri, L., Kohlhaas, K. M., Chen, X., Dikin, D. A., and Ruoff, R. S. (2007). Modulus, fracture strength, and brittle vs. plastic response of the outer shell of arc-grown multi-walled carbon nanotubes, *Exp. Mech.*, **47**, pp. 25–36

238. Barber, A. H., Kaplan-Ashiri, I., Cohen, S. R., Tenne, R., and Wagner, H. D. (2005). Stochastic strength of nanotubes: an appraisal of available data, *Compos. Sci. Technol.*, **65**, pp. 2380–2384.

239. Barber, A. H., Andrews, R., Schadler, L. S., and Wagnera, H. D. (2005). On the tensile strength distribution of multiwalled carbon nanotubes, *Appl. Phys. Lett.*, **87**, pp. 203106_1–203106_3.

240. Ogata, S., and Shibutani, Y. (2003). Ideal tensile strength and band gap of single-walled carbon nanotubes, *Phys. Rev. B*, **68**, pp. 165409_1–165409_4.

241. Ozaki, T., Iwasa, Y., and Mitani, T. (2000). Stiffness of single-walled carbon nanotubes under large strain, *Phys. Rev. Lett.*, **84**, pp. 1712–1715.

242. Dumitrică, T., Belytschko, T., and Yakobson, B. I. (2003). Bond-breaking bifurcation states in carbon nanotube fracture, *J. Chem. Phys.*, **118**(21), pp. 9485–9488.

243. Dumitrica, T., Ming Hua, M., and Yakobson, B. I. (2006). Symmetry-, time-, and temperature-dependent strength of carbon nanotubes, *PNAS*, **103**, pp. 6105–6109.

244. Mielke, S. L., Troya, D., Zhang, S., Li, J.-L., Xiao, S., Car, R., Ruoff, R. S., Schatz, G. C., and Belytschko, T. (2004). The role of vacancy defects and holes in the fracture of carbon nanotubes, *Chem. Phys. Lett.*, **390**, pp. 413–420.

245. Troya, D., Mielke, S. L., and Schatz, G. C. (2003). Carbon nanotube fracture — differences between quantum mechanical mechanisms and those of empirical potentials, *Chem. Phys. Lett.*, **382**, pp. 133–141.

246. Peng, B., Locascio, M., Zapol, P., Li, S., Mielke, S. L., Schatz, G. C., and Espinosa, H. D. (2008). Measurements of near-ultimate strength for multiwalled carbon nanotubes and irradiation-induced crosslinking improvements, *Nat. Nanotech.*, **3**, pp. 626–631.

247. Wen, B., Sader, J. E., and Boland, J. J. (2008). Mechanical Properties of ZnO Nanowires, *Phys. Rev. Lett.*, **101**, pp. 175502_1–175502_4.

248. Hoffmann, S., Utke, I., Moser, B., Michler, J., Christiansen, S. H., Schmidt, V., Senz, S., Werner, P., Gosele, U., and Ballif, C. (2006). Measurement of the bending strength of vapor-liquid-Solid grown silicon nanowires, *Nano Lett.*, **6**, pp. 622–625.

249. Wu, B., Heidelberg, A., Boland, J. J., Sader, J. E., Sun, X. M., and Li, Y. D. (2006). Microstructure-hardened silver nanowires, *Nano Lett.*, **6**, pp. 468–472.

250. Wu, B., Heidelberg, A., and Boland, J. J. (2005). Mechanical properties of ultrahigh-stregth gold nanowires, *Nat. Mater.*, **4**, pp. 525–529.

251. Zhang, S., Mielke, S. L., Khare, R., Troya, D., Ruoff, R. S., Schatz, G. C., and Belytschko, T. (2005). Mechanics of defects in carbon nanotubes: atomistic and multiscale simulations, *Phys. Rev. B*, **71**, pp. 115403_1–115403_12.

252. Ajayan, P. M., Ravikumar, V., Charlier, J.-C. (1998). Surface reconstructions and dimensional changes in single-walled carbon nanotubes, *Phys. Rev. Lett.*, **81**, pp. 1437–1440.

253. Telling, R. H., Ewels, C. P., El-Barbary, A. A., Heggie M. I. (2003). Wigner defects bridge the graphite gap. *Nat. Mater.*, **2**, pp. 333–337.

254. Huang, J. Y., Chen, S., Jo, S. H., Wang, Z., Han, D. X., Chen, G., Dresselhaus, M. S., and Ren, Z. F. (2005). Atomic-scale imaging of wall-by-wall breakdown and concurrent transport measurements in multiwall carbon nanotubes, *Phys. Rev. Lett.*, **94**, pp. 236802_1–236802_4.

255. Huang, J. Y., Chen, S., Wang, Z. Q., Kempa, K., Wang, Y. M., Jo, S. H., Chen, G., Dresselhaus, M. S., and Ren, Z. F. (2006). Superplastic carbon nanotubes, *Nature*, **439**, pp. 281.

256. Troiani, H. E., Miki-Yoshida, M., Camacho-Bragado, G. A., Marques, M. A. L., Rubio, A., Ascencio, J. A., and Jose-Yacaman, M. (2003). Direct observation of the mechanical properties of single-walled carbon nanotubes and their junctions at the atomic level, *Nano Lett.*, **3**, pp. 751–755.

257. von Helden, G., Kemper, P. R., Gotts, N. G., and Bowers, M. T. (1993). Isomers of small carbon cluster anions: linear chains with up to 20 atoms, *Science*, **259**, pp. 1300–1302.

258. Hunter, J., Fye, J., and Jarrold, M. F. (1993). Annealing C_{60}^{+}: synthesis of fullerenes and large carbon rings, *Science*, **260**, pp. 784–786.

259. Rinzler, A. G., Hafner, J. H., Nikolaev, P., Lou, L., Kim, S. G., Tomanek, D., Nordlander, P., Colbert, D. T., and Smalley, R. E. (1995). Unraveling nanotubes: field emission from an atomic wire *Science*, **269**, pp. 1550–1553.

260. Yakobson, B. I., Campbell, M. P., Brabec, C. J., and Bernholc, J. (1997). High strain rate fracture and C-chain unraveling in carbon nanotubes, *Comput. Mater. Sci.*, **8**, pp. 341–348.

261. Xu, C. H. Wang, C. Z., Chan, C. T., and Ho, K. M. (1992). A transferable tight-binding potential for carbon, *J. Phys. Condens. Matter*, **4**, pp. 6047–6054.

262. Van Orden, A., and Saykally, R. J. (1998). Small carbon clusters: spectroscopy, structure, and energetics, *Chem. Rev.*, **98**, pp. 2313–2358.

263. Raghavachari, K., and Binkley, J. S. (1987). Structure, stability, and fragmentation of small carbon clusters, *J. Chem. Phys.*, **87**, pp. 2191–2198.

264. Chen, X., Zhang, S., Dikin, D. A., Ding, W. Ruoff, R. S., Pan, L., and Nakayama, Y. (2003). Mechanics of a carbon nanocoil, *Nano Lett.*, **3**, pp. 1299–1304.

265. Poggi, M. A., Boyles, J. S., Bottomley, L. A., McFarland, A. W., Colton, J. S., Nguyen, C. V., Stevens, R. M., and Lillehei, P. T. (2004). Measuring the compression of a carbon nanospring, *Nano Lett.*, **4**(6), pp. 1009–1016.

266. Gao, P. X., Ding, Y., Mai, W., Hughes, W. L., Lao, C., and Wang, Z. L. (2005). Conversion of zinc oxide nanobelts into superlattice-structured nanohelices, *Science*, **309**, pp. 1700–1704.

267. Singh, J. P., Liu, D.-L., Ye, D.-X., Picu, R. C., Lu, T.-M., and Wang, G.-C. (2004). Metal-coated Si springs: nanoelectromechanical actuators, *Appl. Phys. Lett.*, **84**, pp. 3657–3659.

268. Kong, X. Y., Ding, Y., Yang, R., and Wang, Z. L. (2004). Single-crystal nanorings formed by epitaxial self-coiling of polar nanobelts, *Science*, **303**, pp. 1348–1351.

269. Volodin, A., Buntinx, D., Ahlskog, M., Fonseca, A., Nagy, J. B., and Haesendonck, C. V. (2004). Coiled carbon nanotubes as self-sensing mechanical resonators, *Nano Lett.*, **4**, pp. 1775–1779.

270. Ihara, S., and Itoh, S. (1993). Helically coiled cage forms of graphitic carbon, *Phys. Rev. B*, **48**, pp. 5643–5648.

271. Gayathri, V., Devi, N. R., and Geetha, R. (2010). Hydrogen storage in coiled carbon nanotubes, *Int. J. Hydro. Energy*, **35**, pp. 1313–1320.

272. Zhong-can, O.-Y., Su, Z.-B., and Wang, C.-L. (1997). Coil formation in multishell carbon nanotubes: competition between curvature elasticity and interlayer adhesion, *Phys. Rev. Lett.*, **78**, pp. 4055–4058.

273. Volodin, A., Ahlskog, M., Seynaeve, E., Haesendonck, C. V., Fonseca, A., and Nagy, J. B. (2000). Imaging the elastic properties of coiled carbon nanotubes with atomic force microscopy, *Phys. Rev. Lett.*, **84**, pp. 3342–3345.

274. Ivanov, V., Nagy, J. B., Lambin, Ph., Lucas, A., Zhang, X. B., Zhang, X. F., Bemaerts, D., Tendeloo, G. V., Amelinckx, S., and Van Landuyt, J. (1994). The study of carbon nanotubules produced by catalytic method, *Chem. Phys. Lett.*, **223**, pp. 329–335.

275. Zhang, X. B., Zhang, X. F., Bernaerts, D., Van Tendeloo, G., Amelinckx, S., Van Landuyt, J., Ivanov, V., Nagy, J. B., Lambin, Ph., Lucas, A. A. (1994). The texture of catalytically grown coil-shaped carbon nanotubules, *Europhys. Lett.*, **27**(2), pp. 141–146.

276. Pan, L., Hayashida, T., Zhang, M., and Nakayama, Y. (2001). Field emission properties of carbon tubule nanocoils, *Jpn. J. Appl. Phys.*, **40**, pp. L235–L237.

277. Pradhan, D., and Sharon, M. Carbon nanotubes, nanofilaments and nanobeads by thermal chemical vapor deposition process, *Mater. Sci. Eng. B*, **96**, pp. 24–28.

278. Takenaka, S., Ishida, M., Serizawa, M., Tanabe, E., and Otsuka, K. (2004). Formation of carbon nanofibers and carbon nanotubes through methane decomposition over supported cobalt catalysts, *J. Phys. Chem. B*, **108**, pp. 11464–11472.

279. Xie, J., Mukhopadyay, K., Yadev, J., and Varadan, V. K. (2003). Catalytic chemical vapor deposition synthesis and electron microscopy observation of coiled carbon nanotubes, *Smart Mater. Struct.*, **12**, pp. 744–748.

280. Pan, L., Zhang, M., and Nakayama, Y. (2002). Growth mechanism of carbon nanocoils, *J. Appl. Phys.*, **91**(12), pp. 10058–10061.

281. Koós, A. A., Ehlich, R., Horváth, Z. E., Osváth, Z., Gyulai, J., Nagy, J. B., and Biró, L. P. (2003). STM and AFM investigation of coiled carbon nanotubes produced by laser evaporation of fullerene, *Mater. Sci. Eng. C*, **23**, pp. 275–278.

282. Saveliev, A. V., Merchan-Merchan, W., and Kennedy, L. A. (2003). Metal catalyzed synthesis of carbon nanostructures in an opposed flow methane oxygen flame, *Combust. Flame*, **135**, pp. 27–33.

283. Hokushin, S., Pan, L., and Nakayama, Y. (2007). Diameter control of carbon nanocoils by the catalyst of organic metals, *Jpn. J. Appl. Phys.*, **46**, pp. 5383–5385.

284. Hayashida, T., Pan, L., and Nakayama, Y. (2002). Mechanical and electrical properties of carbon tubule nanocoils, *Physica B*, **323**, pp. 352–353.

285. Biró, L. P., Lazarescu, S. D., Thiry, P. A., Fonseca, A., Nagy, J. B., Lucas, A. A., Lambin, P. (2000). Scanning tunneling microscopy observation

of tightly wound, single-wall coiled carbon nanotubes, *Europhys. Lett.*, **50**, pp. 494–500.

286. Lau, K. T. (2006). Coiled carbon nanotubes: synthesis and their potential applications in advanced composite structures, *Composites B*, **37**, pp. 437–448.

287. Liu, L. Z., Gao, H. L., Zhao, J. J., and Lu, J. P. (2010). Superelasticity of carbon nanocoils from atomistic quantum simulations, *Nanoscale Res. Lett.*, **5** pp. 478–483.

288. Ihara, S., Itoh, S., and Kitakami, J. (1993). Toroidal forms of graphitic carbon, *Phys. Rev. B*, **47**, pp. 12908–12911.

289. Szabó, A., Fonseca, A., Nagy, J. B., Lambin, Ph., Biró, L. P. (2005). Structural origin of coiling in coiled carbon nanotubes, *Carbon*, **43**, pp. 1628–1633.

290. Lu, M., Liu, W. M., Guo, X. Y., and Li, H. L. (2004). Coiled carbon nanotubes growth via reduced-pressure catalytic chemical vapor deposition, *Carbon*, **42**, pp. 805–811.

291. Daraio, C., Nesterenko, V. F., Jin, S., Wang, W., and Rao, A. M. (2004). Impact response by a foamlike forest of coiled carbon nanotubes, *J. Appl. Phys.*, **100**, pp. 064309_1–064309_4.

292. Coluci, V. R., Fonseca, A. F., Galvão and Daraio, C. (2008). Entanglement and the nonlinear elastic behavior of forests of coiled carbon nanotubes, *Phys. Rev. Lett.*, **100**, pp. 086807_1–086807_4.

293. Osváth, Z., Vértesy, G., Tapasztó, L., Wéber, F., Horváth, Z. E., Gyulai, J., and Biró L. P. (2005). Atomically resolved STM images of carbon nanotube defects produced by Ar^+ irradiation. *Phys. Rev. B*, **72**, pp. 045429_1–045429_6.

294. Urita, K., Suenaga, K., Sugai, T., Shinohara, H., and Iijima, S. (2005). *In situ* observation of thermal relaxation of interstitial-vacancy pair defects in a graphite gap. *Phys. Rev. Lett.*, **94**, pp. 155502_1–155502_4.

295. Raghuveer, M. S., Kumar, A., Frederick, M. J., Louie, G. P., Ganesan, P. G., and Ramanath, G. (2006). Site-selective functionalization of carbon nanotubes. *Adv. Mater.*, **18**, pp. 547–552.

296. Jung, Y. J., Homma, Y., Vajtai, R., Kobayashi, Y., Ogino, T., and Ajayan, P. M. (2004). Straightening suspended single walled carbon nanotubes by ion irradiation, *Nano Lett.*, **4**, pp. 1109–1113.

297. Kim, D.-H., Jang, H.-S., Kim, C.-D., Cho, D.-S., Kang, H.-D., and Lee, H.-R. (2003). Enhancement of the field emission of carbon nanotubes straightened by application of argon ion irradiation, *Chem. Phys. Lett.*, **378**, pp. 232–237.

298. Ni, B., Andrews, R., Jacques, D., Qiau, D., Wijesundara, M. B. J., Choi, Y., Hanley, L., and Sinnott, S. B. (2001). A combined computational and experimental study of ion-beam modification of carbon nanotube bundles, *J. Phys. Chem. B*, **105**, pp. 12719–12725.

299. Yang, D. Q., Rochette, J., and Sacher, E. (2005). Controlled chemical functionalization of multiwalled carbon nanotubes, by kiloelectronvolt argon ion treatment and air exposure. *Langmuir*, **21**, pp. 8539–8545.

300. Zhu, Y., Yi, T., Zheng, B., and Cao, L. (1999). The interaction of C_{60} fullerene and carbon nanotube with Ar ion beam, *Appl. Surf. Sci.*, **137**, pp. 83–90.

301. Geim, A. K., and Novoselov, K. S. (2007). The rise of grapheme, *Nature Mater.*, **6**, pp. 183–191.

302. Krasheninnikov, A. V., Nordlund, K., and Keinonen, J. (2002). Production of defects in supported carbon nanotubes under ion irradiation, *Phys. Rev. B*, **65**, pp. 165423_1–165423_8.

303. Lu, A. J., and Pan, B. C. (2004). Nature of single vacancy in achiral carbon nanotubes, *Phys. Rev. Lett.*, **92**, pp. 105504_1–105504_4.

304. Rossato, J., Baierle, R. J., Fazzio, A., and Mota, R. (2005). Vacancy formation process in carbon nanotubes: first-principles approach, *Nano Lett.*, **5**, pp. 197–200.

305. Krasheninnikov, A. V., Lehtinen, P. O., Foster, A. S., and Nieminen, R. M. (2006). Bending the rules: contrasting vacancy energetics and migration in graphite and carbon nanotubes, *Chem. Phys. Lett.*, **418**, pp. 132–136.

306. Ugarte, D. (1992). Curling and closure of graphitic networks under electron-beam irradiation, *Nature*, **359**, pp. 707–709.

307. Banhart, F., and Ajayan, P. M. (1996). Carbon onions as nanoscopic pressure cells for diamond formation, *Nature*, **382**, pp. 433–435.

308. Smith, B. W., Monthioux, M., and Luzzi, D. E. (1998). Encapsulated C_{60} in carbon nanotubes, *Nature*, **396**, pp. 323–324.

309. Gómez-Navarro, C., de Pablo, P., Gómez-Herrero, J., Biel, B., Garcia-Vidal, F., Rubio, A., and Flores, F. (2005). Tuning the conductance of single-walled carbon nanotubes, by ion irradiation in the Anderson localization regime, *Nature Mater.*, **4**, pp. 534–539.

310. Terrones, M., Terrones, H., Banhart, F., Charlier, J.-C., and Ajayan, P. M. (2000). Coalescence of single-walled carbon nanotubes, *Science*, **288**, pp. 1226–1229.

311. Mickelson, W., Aloni, S., Han, W. Q., Cumings, J., and Zettl, A. (2003). Packing C_{60} in boron nitride nanotubes, *Science*, **300**, pp. 467–469.

312. Terrones, M., Banhart, F., Grobert, N., Charlier, J.-C., Terrones, H., and Ajayan, P. M. (2002). Molecular junctions by joining single-walled carbon nanotubes, *Phys. Rev. Lett.*, **89**, pp. 075505_1–075505_4.

313. Skákálova, V., Dettlaff-Weglikowska, U., and Roth, S. (2004). Gamma-irradiated and functionalized single wall nanotubes, *Diamond Relat. Mater.*, **13**, pp. 296–298.

314. Hulman, M., Skákálova, V., Roth, S., and Kuzmany, H. (2005). Raman spectroscopy of single-wall carbon nanotubes and graphite irradiated by γ rays, *J. Appl. Phys.*, **98**, pp. 024311_1–024311_5.

315. Beuneu, F., l'Huillier, C., Salvetat, J. P., Bonard, J. M., and Forró, L. (1999). Modification of multiwall carbon nanotubes by electron irradiation: an ESR study, *Phys. Rev. B*, **59**, pp. 5945–5949.

316. Basiuk, V. A., Kobayashi, K., Kaneko, T., Negishi, Y., Basiuk, E., and Saniger-Blesa, J.-M. (2002). Irradiation of single-walled carbon nanotubes with high-energy protons, *Nano Lett.*, **2**, pp. 789–791.

317. Khare, B., Meyyappan, M., Moore, M. H., Wilhite, P., Imanaka, H., and Chen, B. (2003). Proton irradiation of carbon nanotubes, *Nano Lett.*, **3**, pp. 643–646.

318. Banhart, F. (1999). Irradiation effects in carbon nanostructures, *Rep. Prog. Phys.*, **62**, pp. 1181–1221.

319. Yuzvinsky, T. D., Mickelson, W., Aloni, S., Begtrup, G. E., Kis, A., and Zettl, A. (2006). Shrinking a Carbon Nanotube, *Nano Lett.*, **6**, pp. 2718–2722.

320. Adhikari, A. R., Huang, M. B., Bakhru, H., Talapatra, S., Ajayan, P. M., and Ryu, C. Y. (2006). Effects of proton irradiation on thermal stability of single-walled carbon nanotubes mat, *Nucl. Instrum. Methods Phys. Res. B*, **245**, pp. 431–434.

321. Talapatra, S., Ganesan, P. G., Kim, T., Vajtai, R., Huang, M., Shima, M., Ramanath, G., Srivastava, D., Deevi, S. C., and Ajayan, P. M. (2005). Irradiation-induced magnetism in carbon nanostructures, *Phys. Rev. Lett.*, **95**, pp. 097201_1–097201_4.

322. Tolvanen, A., Kotakoski, J., Krasheninnikov, A. V., and Nordlund, K. (2007). Relative abundance of single and double vacancies in irradiated single-walled carbon nanotubes, *Appl. Phys. Lett.*, **91**, pp. 173109_1–173109_3.

323. O'Rourke Muisener, P. A., Clayton, L., Angelo, J. D., Harmon, J. P., Sukder, A. K., Kumar, A., Cassell, A. M. and Meyyappan, M. (2002). Effects of

gamma radiation on poly(methyl methacrylate)/single-wall nanotube composites, *J. Mater. Res.*, **17**, pp. 2507–2513.

324. Banhart, F. (2006). Irradiation of carbon nanotubes with a focused electron beam in the electron microscope, *J. Mater. Sci.*, **41**, pp. 4505–4511.

325. Baughman, R. H. (2006). Dangerously Seeking Linear Carbon, *Science*, **312**, pp. 1009–1110.

326. Itzhaki, L., Altus, E., Basch, H., and Hoz, S. (2005). Harder than diamond: determining the cross-sectional area and Young's modulus of molecular rods, *Angew. Chem.*, **117**, pp. 7598–7601.

327. Lehtinen, P. O., Foster, A. S., Ma, Y., Krasheninnikov, A. V., and Nieminen, R. M. (2004). Irradiation-induced magnetism in graphite: a density functional study, *Phys. Rev. Lett.*, **93**, pp. 187202_1–187202_4.

328. El-Barbary, A. A., Telling, R. H., Ewels, C. P., Heggie, M. I., and Briddon, P. R. (2003). Structure and energetics of the vacancy in graphite, *Phys. Rev. B*, **68**, pp. 144107_1–144107_7.

329. Warner, J. H., Schäffel, F. G. Zhong, M. H. Rümmeli, B. Büchner, J. Robertson, and G. Briggs, (2009). Investigating the diameter-dependent stability of single-walled carbon nanotubes, *ACS Nano*, **3**, pp. 1557–1563.

330. Niyogi, S., Hamon, M. A., Hu, H., Zhao, B., Bhowmik, P., Sen, R., Itkis, M. E., and Haddon, R. C. (2002). Chemistry of single-walled carbon nanotubes, *Acc. Chem. Res.*, **35**, pp. 1105–1113.

331. Krasheninnikov, A. V., Banhart, F., Li J. X., Foster, A. S., and Nieminen, R. M. (2005). Stability of carbon nanotubes under electron irradiation: role of tube diameter and chirality. *Phys. Rev. B*, **72**, pp. 125428_1–125428_6.

332. Banhart, F., Li, J. X., and Krasheninnikov, A. V. (2005). Carbon nanotubes under electron irradiation: stability of the tubes and their action as pipes for atom transport, *Phys. Rev. B*, **71**, pp. 241408(R)_1–241408(R)_4.

333. Gan, Y., Kotakoski, J., Krasheninnikov, A. V., Nordlun, K., and Banhart, F. (2008). The diffusion of carbon atoms inside carbon nanotubes, *New J. Phys.*, **10**, pp. 023022_1–023022_9.

334. Sun, L., Banhart, F., Krasheninnikov, A. V., Rodríguez-Manzo, J. A., Terrons, M., and Ajayan, P. M. (2006). Carbon nanotubes as high-pressure cylinders and nanoextruders, *Science*, **312**, pp. 1199–1202.

335. Shima, H., Sato, M., Iiboshi, K., Ghosh, S., and Arroyo, M. (2010). Diverse corrugation pattern in radially shrinking carbon nanotubes, *Phys. Rev. B*, **82**, pp. 085401_1–085401_7.

336. Endo, M., Hayashi, T., Muramatsu, H., Kim, Y. A., Terrones, H., Terrones, M., and Dresselhaus, M. S. (2004). Coalescence of double-walled carbon nanotubes: formation of novel carbon bicables, *Nano Lett.*, **4**, pp. 1451–1454.

337. Kawai, T., Yoshiyuki Miyamoto, Y., Sugino, O., and Koga, Y. (2002). General sum rule for chiral index of coalescing ultrathin nanotubes, *Phys. Rev. Lett.*, **89**, pp. 085901_1–085901_4.

338. Li, J. X., and Banhart, F. (2004). The engineering of hot carbon nanotubes with an electron beam, *Nano Lett.*, **4**, pp. 1143–1146.

339. Banhart, F., Li, J. X., and Terrones, M. (2005). Cutting single-walled carbon nanotubes with an electron beam: evidence for atom migration inside nanotubes, *Small*, **1**, pp. 953–956.

340. Jin, C. H., Suenaga, K., Iijima, S. (2008). Plumbing carbon nanotubes, *Nat. Nano.*, **3**, pp. 17–21.

341. Krasheninnikov, A. V., Nordlund, K., Keinonen, J., and Banhart, F. (2003). Making junctions between carbon nanotubes using an ion beam, *Nucl. Instrum. Meth. Phys. Res. B*, **202**, pp. 224–229.

342. Wei, D., and Liu, Y. (2008). The intramolecular junctions of carbon nanotubes, *Adv. Mater.*, **20**, pp. 2815–2841.

343. Jang, I., Sinnott, S. B., Danailov, D., and Keblinski, P. (2004). Molecular dynamics simulation study of carbon nanotube welding under electron beam irradiation, *Nano Lett.*, **4**(1), pp. 109–114.

344. Tsai, P. C., Jeng, Y. R., and Fang, T.-H. (2006). Coalescence, melting, and mechanical characteristics of carbon nanotube junctions, *Phys. Rev. B*, **74**, pp. 045406_1–045406_10.

345. Menon, M., Andriotis, A. N., Srivastava, D., Ponomareva, I., and Chernozatonskii, L. A. (2003). Carbon nanotube "T junctions": formation pathways and conductivity, *Phys. Rev. Lett.*, **91**(14), pp. 145501_1–145501_14.

346. Ponomareva, I., Chernozatonskii, L. A., Andriotis, A. N., and Menon, M. (2003). Formation pathways for single-wall carbon nanotube multiterminal junctions *New J. Phys.*, **5**, pp. 119_1–119_12.

347. Meng, F. Y., Shi, S. Q., Xu, D. S., and Yang, R. (2004). Multiterminal junctions formed by heating ultrathin single-walled carbon nanotubes, *Phys. Rev. B*, **70**, pp. 125418_1–125418_6.

348. Srivastava, D., Menon, M., and Ajayan, P. M. (2003). Branched carbon nanotube junctions predicted by computational nanotechnology, and fabricated through nanowelding, *J. Nanoparticle Res.*, **5**, pp. 395–400.

349. Meng, F. Y., Shi, S. Q., Xu, D. S., and Yang, R. (2003). Size effect of X-shaped carbon nanotube junctions, *Carbon*, **44**, pp. 1263–1266.

350. Zhao, Y., Smalley, R. E., and Yakobson, B. I. (2002). Coalescence of fullerene cages: topology, energetics, and molecular dynamics simulation, *Phys. Rev. B*, **66**, pp. 195409_1–195409_9.

351. Cumings, J., and Zettl, A. (2000). Low-friction nanoscale linear bearing realized from multiwall carbon nanotubes, *Science*, **289**, pp. 602–604.

352. Yu, M.-F., Yakobson, B. I., and Ruoff, R. S. (2000). Controlled sliding and pullout of nested shells in individual multiwalled carbon nanotubes, *J. Phys. Chem. B*, **104**, pp. 8764–8767.

353. Bourlon, B., Glattli, D. C., Miko, C., Forro, L., and Bachtold, A. (2004). Carbon nanotube based bearing for rotational motions, *Nano Lett.*, **4**, pp. 709–712.

354. Deshpande, V. V., Chiu, H.-Y., Postma, H. W. Ch., Mikó, C., Forró, L., and Bockrath, M. (2006). Carbon nanotube linear bearing nanoswitches, *Nano Lett.*, **6**, pp. 1092–1095.

355. Subramanian, A., Dong, L. X., Nelson, B. J., and Ferreira, A. (2010). Supermolecular switches based on multiwalled carbon nanotubes, *Appl. Phys. Lett.*, **96**, pp. 073116_1–073116_3.

356. Fennimore, A. M., Yuzvinsky, T. D., Han, W.-Q., Fuhrer, M. S., Cumings, J., and Zettl, A. (2003). Rotational actuators based on carbon nanotubes, *Nature*, **424**, pp. 408–410.

357. Bourlon, B., Glattli, D. C., Plaçais, B., Berroir, J. M., Forró, L., and Bachtold, A. (2004). Geometrical dependence of high-bias current in multiwalled carbon nanotubes, *Phys. Rev. Lett.*, **92**, pp. 026804_1–026804_4.

358. Dong, L., Nelson, B. J., Fukuda, T., and Arai, F. (2006). Towards nanotube linear servomotors, *IEEE Trans.*, **3**(3), pp. 228–235.

359. Barreiro, A., Rurali, R., Hernández, E. R., Moser, J., Pichler, T., Forró, L., and Bachtold, A. (2008). Subnanometer motion of cargoes driven by thermal gradients along carbon nanotubes, *Science*, **320**, pp. 775–778.

360. Servantie, J., and Gaspard, P. (2006). Rotational dynamics and friction in double-walled carbon nanotubes, *Phys. Rev. Lett.*, **97**, pp. 186106_1–186106_4.

361. Maslov, L. (2006). Concept of nonvolatile memory based on multiwall carbon nanotubes, *Nanotechnology*, **17**, pp. 2475–2482.

362. Subramanian, A., Dong, L., and Nelson, B. J. (2009). Stability and analysis of configuration-tunable bi-directional MWNT bearings, *Nanotechnology*, **20**, pp. 495704_1–495704_7.

363. Takagi, Y., Uda, T., and Ohno, T. (2008). Carbon nanotube motors driven by carbon nanotube, *J. Chem. Phys.*, **128**, pp. 194704_1–194704_7.

364. Yan, Q., Zhou, G., Hao, S., Wu, J., and Duan, W. (2006). Mechanism of nanoelectronic switch based on telescoping carbon nanotubes, *Appl. Phys. Lett.*, **88**, pp. 173107_1–173107_3.

365. Bailey, S. W. D., Amanatidis, I., and Lambert, C. J. (2008). Carbon nanotube electron windmills: a novel design for nanomotors, *Phys. Rev. Lett.*, **100**, pp. 256802_1–256802_4.

366. Li, Y., Hu, N., Yamamoto, G., Wang, Z., Hashida, T., Asanuma, H., Dong, C., Okabe, T., Arai, M., and Fukunaga, H. (2010). Molecular mechanics simulation of the sliding behavior between nested walls in a multi-walled carbon nanotube, *Carbon*, **48**, pp. 2934–2940.

367. Neild, A., Ng, T. W., and Zheng, Q. (2009). Controlled driven oscillations of double-walled carbon nanotubes, *Eur. Phys. Lett.*, **87**, pp. 16002_1–16002_5.

368. Po, G., and Ghoniem, N. M. (2010). Coupled oscillations of double-walled carbon nanotubes, *J. Appl. Phys.*, **107**, pp. 094310_1–094310_5.

369. Charlier, J.-C., and Michenaud, J.-P. (1993). Energetics of multilayered carbon tubules, *Phys. Rev. Lett.*, **70**, pp. 1858–1861.

370. Dienwiebel, M., Verhoeven, G. S., Pradeep, N., Frenken, J. W. M., Heimberg, J. A., and Zandbergen, H. W. (2004). Superlubricity of graphite., *Phys. Rev. Lett.*, **92**, pp. 126101_1–126101_4.

371. Kolmogorov, A. N., and Crespi, V. H. (2000). Smoothest bearings: interlayer sliding in multiwalled carbon nanotubes, *Phys. Rev. Lett.*, **85**, pp. 4727–4730.

372. Saito, R., Matsuo, R., Kimura, T., Dresselhaus, G., and Dresselhaus, M. S. (2001). Anomalous potential barrier of double-wall carbon nanotube, *Chem. Phys. Lett.*, **348**, pp. 187–193.

373. Damnjanović, M., Vuković, T., and Milošević, I. (2002). Super-slippery carbon nanotubes, *Eur. Phys. J. B*, **25**, pp. 131–134.

374. Xia, Z., and Curtin, W. A. (2004). Pullout forces and friction in multiwall carbon nanotubes, *Phys. Rev. B*, **69**, pp. 233408_1–233408_4.

375. Zhao, Y., Ma, C.-C., Chen, G. H., and Jiang, Q. (2003). Energy dissipation mechanisms in carbon nanotube oscillators. *Phys. Rev. Lett.*, **91**, pp. 175504_1–175504_4.

376. Tangney, P, Louie, S. G., and Cohen, M. L. (2004). Dynamic sliding friction between concentric carbon nanotubes, *Phys. Rev. Lett.*, **93**(6), pp. 065503_1–065503_4.

377. Servantie, J., and Gaspard, P. (2003). Methods of calculation of a friction coefficient: application to nanotubes, *Phys. Rev. Lett.*, **91**, pp. 185503_1–185503_4.

378. Socoliuc, A., Bennewitz, R. Gnecco, E., and Meyer, E. (2004). Transition from stick-slip to continuous sliding in atomic friction: entering a new regime of ultralow friction, *Phys. Rev. Lett.*, **92**, pp. 134301_1–134301_4

379. Zheng, Q., and Jiang, Q. (2002). Multiwalled carbon nanotubes as gigahertz oscillators, *Phys. Rev. Lett.*, **88**(4), pp. 045503_1–045503_4.

380. Rivera, J. L., McCabe, C., and Cummings, P. T. (2003). Oscillatory behavior of double-walled nanotubes under extension: a simple nanoscale damped spring, *Nano Lett.*, **3**, pp. 1001–1005.

381. Hong, B. H., Small, J. P., Purewal, M. S., Mullokandov, A., Sfeir, M. Y., Wang, F., Lee, J. Y., Heinz, T. F., Brus, L. E., Kim, P., and Kim, K. S. (2005). Extracting subnanometer single shells from ultralong multiwalled carbon nanotubes, *Proc. Natl. Acad. Sci.*, **102**, pp. 14155–14158.

382. Hertel, T., Walkup, R. E., and Avouris, P. (1998). Deformation of carbon nanotubes by surface van der Waals forces, *Phys. Rev. B*, **58**, pp. 13870–13873.

383. Williams, P. A., Papadakis, S. J., Patel, A. M., Falvo, M. R., Washburn, S., and Superfine, R. (2002). Torsional response and stiffening of individual multiwalled carbon nanotubes, *Phys. Rev. Lett.*, **89**(25), pp. 255202_1–255202_4.

384. Williams, P. A., Papadakis, S. J., Patel, A. M., Falvo, M. R., Washburn, S., and Superfine, R. (2003). Fabrication of nanometer-scale mechanical devices incorporating multiwalled carbon nanotubes as torsional springs, *Appl. Phys. Lett.*, **82**, pp. 805–807.

385. Papadakis, S. J., Hall, A. R., Williams, P. A., Vicci, L., Falvo, M. R., Superfine, R., and Washburn, S. (2004). Resonant oscillators with carbon-nanotube torsion springs, *Phys. Rev. Lett.*, **93**(14), pp. 146101_1–146101_4.

386. Collins, P. G., Hersam, M., Arnold, M., Martel, R., and Avouris, P. (2001). Current saturation and electrical breakdown in multiwalled carbon nanotubes, *Phys. Rev. Lett.*, **86**, pp. 3128–3131.

387. Collins, P. G., Arnold, M. S., and Avouris, P. (2001). Engineering carbon nanotubes and nanotube circuits using electrical breakdown, *Science*, **292**, pp. 706–709.

388. Soule, D. E., and Nezbeda, C. W. (1968). Direct basal-plane shear in single-crystal graphite, *J. Appl. Phys.*, **39**, pp. 5122–5139.

389. Zheng, Q., Liu, J. Z., and Jiang, Q. (2002). Excess van der Waals interaction energy of a multiwalled carbon nanotube with an extruded core and the induced core oscillation, *Phys. Rev. B*, **65** pp. 245409_1–245409_6.

390. Henrard, L., Hernandez, E., Bernier, P., and Rubio, A. (1999). van der Waals interaction in nanotube bundles: consequences on vibrational modes, *Phys. Rev. B*, **60**, pp. R8521–R8524.

391. Kang, J. W., and Lee, J. H. (2008). Frequency characteristics of triple-walled carbon nanotube gigahertz devices, *Nanotechnology*, **19**, pp. 285704_1–285704_6.

392. Lebedeva, I. V., Knizhnik, A. A., Popov, A. M., Lozovik, Y. E., and Potapkin, B. V. (2009). Dissipation and fluctuations in nanoelectromechanical systems based on carbon nanotubes, *Nanotechnology*, **20**, pp. 105202_1–105202_13.

393. Ershova, O. V., Lebedeva, I. V., Lozovik, Y. E., Popov, A. M., Knizhnik, A. A., Potapkin, B. V., Bubel, O. N., Kislyakov, E. F., and Poklonskii, N. A. (2010). Nanotube-based nanoelectromechanical systems: control versus thermodynamic fluctuations, *Phys. Rev. B*, **81**, pp. 155453_1–155453_15.

394. Jensen, K., Girit, C., Mickelson, W., and Zettl, A. (2006). Tunable nanoresonators constructed from telescoping nanotubes, *Phys. Rev. Lett.*, **96**, pp. 215503_1–215503_4.

395. Cleland, A. N., and Roukes, M. L. (1996). Fabrication of high frequency nanometer scale mechanical resonators from bulk Si crystals *Appl. Phys. Lett.*, **69**, pp. 2653–2625.

396. Dalton, A. B., Collins, S., Munoz, E., Razal, J. M., Ebron, V. H., Ferraris, J. P., Coleman, J. N., Kim, B. G., and Baughman, R. H. (2003). Super-tough carbon-nanotube fibres, Nature (London) **423**, pp. 703.

397. Yang, L., Park, C.-H., Son, Y.-W., Cohen, M. L., and Louie, S. G. (2007). Quasiparticle energies and band gaps in graphene nanoribbons, *Phys. Rev. Lett.*, **99**, pp. 186801_1–186801_4.

398. Chen, Z. H., Lin, Y. M., Rooks, M. J., and Avouris, P. (2007). Graphene nano-ribbon electronics, *Physica E*, **40**, pp. 228–232.

399. Han, M. Y., Ozyilmaz, B., Zhang, Y. B., and Kim, P. (2007). Energy band-gap engineering of graphene nanoribbons, *Phys. Rev. Lett.*, **98**, pp. 206805_1–206805_4.

400. Tapaszto, L., Dobrik, G., Lambin, P., and Biro, L. P. (2008). Tailoring the atomic structure of graphene nanoribbons by scanning tunnelling microscope lithography, *Nat. Nanotechnol.*, **3**, pp. 397–401.

401. Bai, J. W., Duan, X. F., and Huang, Y. (2009). Rational fabrication of graphene nanoribbons using a nanowire etch mask, *Nano Lett.*, **9**, pp. 2083–2087.

402. Wang, X. R., and Dai, H. J. (2010). Etching and narrowing of graphene from the edges, *Nat. Chem.*, **2**, pp. 661–665.

403. Li, X. L., Wang, X. R., Zhang, L., Lee, S. W., and Dai, H. J. (2008). Chemically derived, ultrasmooth graphene nanoribbon semiconductors, *Science*, **319**, pp. 1229–1232.

404. Wu, Z. S., Ren, W. C., Gao, L. B., Liu, B. L., Zhao, J. P., and Cheng, H. M. (2010). Efficient synthesis of graphene nanoribbons sonochemically cut from graphene sheets, *Nano Res.*, **3**, pp. 16–22.

405. Campos-Delgado, J., Romo-Herrera, J. M., Jia, X. T., Cullen, D. A., Muramatsu, H., Kim, Y. A., Hayashi, T., Ren, Z. F., Smith, D. J., Okuno, Y., Ohba, T., Kanoh, H., Kaneko, K., Endo, M., Terrones, H., Dresselhaus, M. S., and Terrones, M. (2008). Bulk production of a new form of sp^2 carbon: crystalline graphene nanoribbons, *Nano Lett.*, **8**, pp. 2773–2778.

406. Wei, D. C., Liu, Y. Q., Zhang, H. L., Huang, L. P., Wu, B., Chen, J. Y., and Yu, G. (2009). Scalable synthesis of few-layer graphene ribbons with controlled morphologies by a template method and their applications in nanoelectromechanical switches, *J. Am. Chem. Soc.*, **131**, pp. 11147–11154.

407. Datta, S. S., Strachan, D. R., Khamis, S. M., and Johnson, A. T. C. (2008). Crystallographic etching of few-layer graphene, *Nano Lett.*, **8**(7), pp. 1912–1915.

408. Ci, L. J., Xu, Z. P., Wang, L. L., Gao, W., Ding, F., Kelly, K. F., Yakobson, B. I., and Ajayan, P. M. (2008). Controlled nanocutting of graphene, *Nano Res.*, **1**, pp. 116–122.

409. Campos, L. C., Manfrinato, V. R., Sanchez-Yamagishi, J. D., Kong, J., and Jarillo-Herrero, P. (2009). Anisotropic etching and nanoribbon formation in single-layer graphene, *Nano Lett.*, **9**, pp. 2600–2604.

410. Kosynkin, D. V., Higginbotham, A. L., Sinitskii, A., Lomeda, J. R., Dimiev, A., Price, B. K., and Tour, J. M. (2009). Longitudinal unzipping of carbon nanotubes to form graphene nanoribbons, *Nature*, **458**, pp. 872–876.

411. Santos, H., Chico, L., and Brey, L. (2009). Carbon nanoelectronics: unzipping tubes into graphene ribbons, *Phys. Rev. Lett.*, **103**, pp. 086801_1–086801_4.

412. Jiao, L. Zhang, L., Wang, X., Diankov, G., and Dai, H. (2009). Narrow graphene nanoribbons from carbon nanotubes, *Nature* **458**, pp. 877–880.

413. Cano-Márquez, A. G., Rodriguez-Macías, F. J., Campos Delgado, J., Espinosa-González, C. G., Tristán-López, F, Ramírez-Gonzalez, C. G., Cullen, D. A., Smith, D. J, Terrones, M., and Vega-Cantu, Y. I. (2009). Ex-MWNTs: graphene sheets and ribbons produced by lithium intercalation and exfoliation of carbon nanotubes, *Nano Lett.*, **9**(4), pp. 1527–1533.

414. Elías, A. L., Botello-Méndez, A. R., Meneses-Rodríguez, D., González, V. J., Ramírez-González, D., Ci, L., Muñoz-Sandoval, E, Ajayan, P. M., Terrones, H, and Terrones, M. (2010). Longitudinal cutting of pure and doped carbon nanotubes to form graphitic nanoribbons using metal clusters as nanoscalpels, *Nano Lett.*, **10**, pp. 366–372.

415. Kim, K., Sussman, A., and Zettl, A. (2010). Graphene nanoribbons obtained by electrically unwrapping carbon nanotubes, *ACS Nano* **4**(3), pp. 1362–1366.

416. Jiao, L., Zhang, L., Ding, L., Liu, J., and Dai, H. (2010). Aligned graphene nanoribbons and crossbars from unzipped. Carbon nanotubes, *Nano Res.*, **3**, pp. 387–394.

417. Liang, F., Sadana, A. K., Peera, A., Chattopadhyay, J., Gu, Z., Hauge, R. H., and Billups, W. E. (2004). A convenient route to functionalized carbon nanotubes, *Nano Lett.*, **4**, pp. 1257–1260.

418. Chattopadhyay, J., Sadana, A. K., Liang, F., Beach, J. M., Xiao, Y., Hauge, R. H., and Billups, W. E. (2005). Carbon nanotube salts. Arylation of single-wall carbon nanotubes, *Org. Lett.*, **7**(19), pp. 4067–4069.

419. Ci, L., Xu, Z., Wang, L., Gao, W., Ding, F., Kelly, K. F., Yakobson, B. I., and Ajayan, P. M. (2008). Controlled nanocutting of graphene, *Nano Res.*, **1**, pp. 116–122.

420. Yuzvinsky, T. D., Mickelson, W., Aloni, S., Konsek, S. L., Fennimore, A. M., Begtrup, G. E., Kis, A., Regan, B. C., and Zettl, A. (2005). Imaging the life story of nanotube devices, *Appl. Phys. Lett.*, **87**, pp. 053109_1–053109_3.

421. Kim, W. S., Moon, S. Y., Bang, S. Y., Choi, B. G., Ham, H., Sekino, T., and Shim, K. B. Fabrication of graphene layers from multiwalled carbon nanotubes using high dc pulse. *Appl. Phys. Lett.*, **95**, pp. 083103_1–083103_3.

422. Ajayan, P. M., and Tour, J. M. (2007). Nanotube composites, *Nature*, **447**, pp. 1066–1068.

423. Mehta, P., and Monteiro, P. (2005). *Concrete: Microstructure, Properties, and Materials, 3rd ed.* (McGraw-Hill Professional).

424. Bard, K. A. (2008). *An Introduction to the Archaeology of Ancient Egypt* (Oxford: Blackwell Publishing Ltd).

425. Hata, K., Futaba, D. N., Mizuno, K., Namai, T., Yumura, M., and Iijima, S. (2004). Water-assisted highly efficient synthesis of impurity-free singlewalled carbon nanotubes. *Science*, **306**, pp. 1362–1364.

426. Chou, T.-W. (1992). *Microstructural Design of Fiber Composites* (Cambridge University Press).

427. Frankland, S. J. V., Caglar, A., Brenner, D. W., and Griebel, M. (2002). Molecular simulation of the influence of chemical cross-links on the shear strength of carbon nanotube–polymer interfaces, *J. Phys. Chem. B*, **106**, pp. 3046–3048.

428. Liao, K., and Li, S. (2001). Interfacial characteristics of a carbon nanotube-polystyrene composite system, *Appl. Phys. Lett.*, **79**, pp. 4225–4227.

429. Frogley, M. D., Ravich, D., and Wagner, H. D. (2003). Mechanical properties of carbon nanoparticle-reinforced elastomers, *Compos. Sci. Technol.*, **63**, pp. 1647–1654.

430. Dai, H. (2002). Carbon nanotubes: opportunities and challenges, *Surf. Sci.*, **500**, pp. 218–241.

431. Harris, P. J. F. (2004). Carbon nanotube composites. *Int. Mater. Rev.*, **49**, pp. 31–43.

432. Jin, L., Bower, C., and Zhou, O. (1999). Alignment of carbon nanotubes in a polymer matrix by mechanical stretching, *Appl. Phys. Lett.*, **73**(9), pp. 1197–1199.

433. Haggenmueller, R., Gommans, H. H., Rinzler, A. G., Fischer, J. E., and Winey, K. I. (2000). Aligned single-wall carbon nanotubes composites by melt processing methods, *Chem. Phys. Lett.*, **330**(3–4), pp. 219–225.

434. Shaffer, M., and Kinloch, I. A. (2004). Prospects for nanotube and nanofibre composites, *Compos. Sci. Technol.*, **64**, pp. 2281–2282.

435. Tucker, C. L., and Liang, E. (1999). Stiffness predictions for unidirectional shortfiber composites: review and evaluation, *Compos. Sci. Technol.*, **59**. 5, pp. 655–671.

436. Daniel, I. M., and Ishai, O. (1994). *Engineering Mechanics of Composite Materials*, (Oxford University Press).

437. Callister, W. D. (2003). *Materials Science and Engineering, an Introduction*, (Wiley, New York).

438. Cox, H. L. (1952). The elasticity and strength of paper and other fibrous materials, *Br. J. Appl. Phys.*, **3**, pp. 72–79.

439. Kelley, A. (1966). *Strong Solids*, (Oxford University Press).

440. Zhang, X., Cao, A., Wei, B., Li, Y., Wei, J., Xu, C., Wu, D. (2002). Rapid growth of well aligned carbon nanotube arrays, *Chem. Phys. Lett.*, **362**, pp. 285–290.

441. Zhong, G., Iwasaki, T., Robertson, J., and Kawarada, H. (2007). Growth kinetics of 0.5 cm vertically aligned single-walled carbon nanotubes, *J. Phys. Chem. B*, **111**, pp. 1907–1910.

442. Ci, L., Suhr, J., Pushparaj, V., Zhang, X., and Ajayan, P. M. (2008). Continuous carbon nanotube reinforced composites, *Nano Lett.*, **8**(9), pp. 2762–2766.

443. Yap, H. W., Lakes, R. S., and Carpick, R. W. (2007). Mechanical instabilities of individual multiwalled carbon nanotubes under cyclic axial compression, *Nano Lett.*, **7**, pp. 1149–1154.

444. Thostenson, E. T., and Chou, T.-W. (2004). Nanotube buckling in aligned multi-wall carbon nanotube composites, *Carbon*, **42**, pp. 3015–3018.

445. Kuzumaki, T., and Mitsuda, Y. (2006). Nanoscale mechanics of carbon nanotube evaluated by nanoprobe manipulation in transmission electron Microscope, *Jpn. J. Appl. Phys.*, **45**, pp. 364–368.

446. Schadler, L. S., Giannaris, S. C and Ajayan, P. M. (1998). Load transfer in carbon nanotube epoxy composites, *Appl. Phys. Lett.*, **73**(26), pp. 3842–3844.

447. Andrews, R., Jacques, D., Rao, A. M., Rantell, T., Derbyshire, F., Chen, Y., Chen, J., and Haddon, R. C. (1999). Nanotube composite carbon fibers, *Appl. Phys. Lett.*, **75**, pp. 1329–1331.

448. Mamedov, A. A., Kotov, N. A., Prato, M., Guldi, D. M., Wicksted, J. P., and Hirsch, A. (2002). Molecular design of strong single-wall carbon nanotube/polyelectrolyte multilayer composites, *Nature Mater.*, **1**, pp. 190–194.

449. Frogley, M. D., Ravich, D., and Wagner, H. D. (2003). Mechanical properties of carbon nanoparticle-reinforced elastomers, *Compos. Sci. Technol.*, **63**, pp. 1647–1654.

450. Barber, A. H., Cohen, S. R., and Wagner, H. D. (2003). Measurement of carbon nanotube–polymer interfacial strength, *Appl. Phys. Lett.*, **82**, pp. 4140–4142.

451. Wagner, H. D., and Vaia, R. A. (2004). Carbon nanotube-based polymer composites: outstanding issues at the interface for mechanics, *Mater. Today*, **7**, pp. 38–42.

452. Barber, A. H., Cohen, S. R., Kenig, S., and Wagner, H. D. (2004). Interfacial fracture energy measurements for multi-walled carbon nanotubes pulled from a polymer matrix, *Compos. Sci. Technol.*, **64**, pp. 2283–2289.

453. Liu, J. Q., Xiao, T., Liao, K., and Wu, P. (2007). Interfacial design of carbon nanotube polymer composites: a hybrid system of noncovalent and covalent functionalizations, *Nanotechnology*, **18**, pp. 165701_1–165701_6.

454. Kim, B. W., and Nairn, J. A. (2002). Observations of fiber fracture and interfacial debonding phenomena using the fragmentation test in single fiber composites, *J. Comp. Mater.*, **36**, pp. 1825–1858.

455. Zhandarov, S., Pisanova, E., Mader, E., and Nairn, J. A. (2001). Investigation of load transfer between the fiber and the matrix in pull-out tests with fibers having different diameters, *J. Adhesion Sci. Tech.*, **15**, pp. 205–222.

456. Pisanova, E., Zhandarov, S., Mader, E., Ahmad, I., and Young, R. J. (2001). Three techniques of interfacial bond strength estimation from direct observation of crack initiation and propagation in polymer-fibre systems, *Composites A*, **32**, pp. 435–443.

457. Pisanova, E., Zhandarov, S., and Mader, E. (2001). How can adhesion be determined from micromechanical tests? *Composites A*, **32**, pp. 425–434.

458. Nairn, J. A. (2000). Analytical fracture mechanics analysis of the pull-out test including the effects of friction and thermal stresses, *Adv. Comp. Lett.*, **9**, pp. 373–383.

459. Leung, C. K. Y., and Li, V. C. (1991). New strength-based model for the debonding of discontinuous fibres in an elastic matrix, *J. Mater. Sci.*, **26**, pp. 5996–6010.

460. Kerans, R. J., and Parthasarathy, T. A. (19911). Theoretical Analysis of the Fiber Pullout and Pushout Tests, *J. Am. Ceramic Soc.*, **74**, pp. 1585–1596.

461. Jia, Z., Wang, Z., Xu, C., Liang, J., Wei, B., Wu, D., and Zhu, S. (1999). Study on poly(methyl methacrylate)/carbon nanotube composites, *Mater. Sci. Eng. A*, **271**, pp. 395–400.

462. Mylvaganam, K., and Zhang, L. C. (2004). Chemical bonding in polyethylene-nanotube composites: a quantum mechanics prediction, *J. Phys. Chem. B*, **108**, pp. 5217–5220.

463. Eitan, A., Jiang, K., Dukes, D., Andrews, R., and Schadler, L. S. (2003). Surface modification of multiwalled carbon nanotubes: toward the tailoring of the interface in polymer composites, *Chem. Mater.*, **15**, pp. 3198–3201.

464. Goh, H. W., Goh, S. H., Xu, G. Q., Pramoda, K. P., and Zhang, W. D. (2003). Dynamic mechanical behavior of *in situ* functionalized multi-walled carbon nanotube/phenoxy resin composite, *Chem. Phys. Lett.*, **373**, pp. 277–283.

465. Jang, J., Bae, J., and Yoon, S.-H. (2003). A study on the effect of surface treatment of carbon nanotubes for liquid crystalline epoxide-carbon nanotube composites, *J. Mater. Chem.*, **13**, pp. 676–681.

466. Sun, Y.-P., Fu, K., Lin, Y., and Huang, W. (2002). Functionalized carbon nanotubes: properties and applications, *Acc. Chem. Res.*, **35**, pp. 1096–1104.

467. Wong, M., Paramsothy, M., Xu, X. J., Ren, Y., Li, S., and Liao, K. (2003). Physical interactions at carbon nanotube–polymer interface, *Polymer*, **44**, pp. 7757–7764.

468. Bai, J. B., and Allaoui, A. (2003). Effect of the length and the aggregate size of MWNTs on the improvement efficiency of the mechanical and electrical properties, *Composites A*, **34**, pp. 689–694.

469. Bai, J. B. (2003). Evidence of the reinforcement role of CVD multi-walled carbon nanotubes in a polymer matrix, *Carbon*, **41**, pp. 1331–1334.

470. Wagner, H. D., Lourie, O., Feldman, Y., and Tenne, R. (1998). Stress-induced fragmentation of multiwall carbon nanotubes in a polymer matrix, *Appl. Phys. Lett.*, **72**, pp. 188–190.

471. Lordi, V., and Yao, N. (2002). Molecular mechanics of binding in carbon nanotubes-polymer composites, *J. Mater. Res.*, **15**, pp. 2770–2779.

472. Moon, C. K., Lee, J.-O., Cho, H. H., and Kim, K. S. (1992). Effect of diameter and surface treatment of fiber on interfacial shear strength in glass fiber/epoxy and HDPE, *J. Appl. Polym. Sci.*, **45**, pp. 443–450.

473. Koss, D. A., Petrich, R. R., Kallas, M. N., and Hellmann, J. R. (1994). Interfacial shear and matrix plasticity during fiber push-out in a metal-matrix composite, *Compos. Sci. Technol.*, **51**, pp. 27–33.

474. Tripathi, D., Turton, T., Chen, F., and Jones, F. (1997). A new method to normalize the effect of matrix properties on the value of interfacial shear strength obtained from the fragmentation test, *J. Mater. Sci.*, **32**, pp. 4759–4765.

475. Huang, Y., and Young, R. J. (1996). Interfacial micromechanics in thermoplastic and thermosetting matrix carbon fibre composites, *Composites A*, **27**, pp. 973–980.

476. Chua, P. S., and Piggott, M. R. (1985) The glass fibre–polymer interface: I—theoretical consideration for single fibre pull-out tests, *Compos. Sci. Technol.*, **22**, pp. 33–42.

477. Ajayan, P. M., Stephan, O., Colliex, C., and Trauth, D. (1994). Aligned carbon nanotube arrays formed by cutting a polymer resin-nanotube composite, *Science*, **265**(5176), pp. 1212–1214.

478. Cooper, C. A., Cohen, S. R., Barber, A. H., and Wagner, H. D. (2002). Detachment of nanotubes from a polymer matrix, *Appl. Phys. Lett.*, **81**(20), pp. 3873–3875.

479. Coleman, J. N., Khan, U., Blau, W. J., and Gun'ko, Y. K. (2006). Small but strong: a review of the mechanical properties of carbon nanotube–polymer composites, *Carbon*, **44**, pp. 1624–1652.

480. Dufresne, A., Paillet, M., Putaux, J. L., Canet, R., Carmona, F., Delhaes, P., and Cui, S. (2002). Processing and characterization of carbon nanotube/poly(styrene-co-butyl acrylate) nanocomposites, *J. Mater. Sci.*, **37**(18), pp. 3915–3923.

481. Probst, O., Moore, E. M., Resasco, D. E., and Grady, B. P. (2004). Nucleation of polyvinyl alcohol crystallization by single-walled carbon nanotubes, *Polymer*, **45**(13), pp. 4437–4443.

482. Dalmas, F., Chazeau, L., Gauthier, C., Masenelli-Varlot, K., Dendievel, R., Cavaillé, J. Y., and Folló, L. et al. (2005). Multiwalled carbon nanotube/polymer nanocomposites: processing and properties, *J. Polym. Sci. Part B: Polym. Phys.*, **43**, pp. 1186–1197.

483. Shaffer, M. S. P., and Windle, A. H. (1999). Fabrication and characterization of carbon nanotube/poly(vinyl alcohol) composites, *Adv. Mater.*, **11**(11), pp. 937–941.

484. Coleman, J. N., Cadek, M., Blake, R., Nicolosi, V., Ryan, K. P., Belton, C., Fonseca, A., Nagy, J. B., Gun'ko, Y. K., and Blau, W. J. (2004). High performance nanotube-reinforced plastics: understanding the mechanism of strength increase, *Adv. Funct. Mater.*, **14**(8), pp. 791–798.

485. Andrews, R., Jacques, D., Minot, M., and Rantell, T. (2002). Fabrication of carbon multiwall nanotube/polymer composites by shear mixing, *Macromol. Mater. Eng.*, **287**(6), pp. 395–403.

486. Breuer, O., and Sundararaj, U. (2004). Big returns from small fibers: a review of polymer/carbon nanotube composites, *Polym. Compos.*, **25**(6), pp. 630–645.

487. Jin, Z., Pramoda, K., Xu, G., and Goh, S. H. (2001). Dynamic mechanical behavior of melt-processed multi-walled carbon nanotube/poly (methyl methacrylate) composites, *Chem. Phys. Lett.*, **337**(1–3), pp. 43–47.

488. Liu, T. X., Phang, I. Y., Shen, L., Chow, S. Y., and Zhang, W.-D. (2004). Morphology and mechanical properties of multiwalled carbon nanotubes reinforced nylon-6 composites, *Macromolecules*, **37**(19), pp. 7214–7222.

489. De Zhang, W., Shen, L., Phang, I. Y., and Liu, T. X. (2004). Carbon nanotubes reinforced nylon-6 composite prepared by simple melt-compounding, *Macromolecules*, **37**(2), pp. 256–259.

490. Ajayan, P. M., Schadler, L. S., Giannaris, C., and Rubio, A. (2000). Single-walled carbon nanotube–polymer composites: strength and weakness, *Adv. Mater.*, **12**(10), pp. 750–753.

491. Xu, X., Thwe, M. M., Shearwood, C., and Liao, K. (2002). Mechanical properties and interfacial characteristics of carbon-nanotube-reinforced epoxy thin films, *Appl. Phys. Lett.*, **81**(15), pp. 2833–2835.

492. Allaoui, A., Bai, S., Cheng, H. M., and Bai, J. B. (2002). Mechanical and electrical properties of a MWNT/epoxy composite, *Compos. Sci. Technol.*, **62**(15), pp. 1993–1998.

493. Bai, J. (2003). Evidence of the reinforcement role of chemical vapour deposition multi-walled carbon nanotubes in a polymer matrix, *Carbon*, **41**(6), pp. 1325–1328.

494. Viswanathan, G., Chakrapani, N., Yang, H., Wei, B., Chung, H., Cho, K., Ryu, C. Y., and Ajayan, P. M. (2003). Single-step *in situ* synthesis of polymer-grafted single-wall nanotube composites, *J. Am. Chem. Soc.*, **125**(31), pp. 9258–9259.

495. Fu, K., Huang, W., Lin, Y., Riddle, L. A., Carroll, D. L., and Sun, Y. P. (2001). Defunctionalization of functionalized carbon nanotubes, *Nano Lett.*, **1**(8), pp. 439–441.

496. Peigney, A. (2002). Tougher ceramics with nanotubes, *Nat. Mater.*, **2**, pp. 15–16.

497. Cho, J., Boccaccini, A. R., and Shaffer, M. S. P. (2009). Ceramic matrix composites containing carbon nanotubes, *J. Mater. Sci.*, **44**, pp. 1934–1951.

498. Sun, J., Gao, L., and Li, W. (2002). Colloidal processing of carbon nanotube/alumina composites, *Chem. Mater.*, **14**, pp. 5169–5172.

499. Zhan, G. D., Kuntz, J. D., Wan, J. L., and Mukherjee, A. K. (2003). Single-wall carbon nanotubes as attractive toughening agents in alumina-based nanocomposites, *Nat. Mater.*, **2**, pp. 38–42.

500. Wang, X. T., Padture, N. P., and Tanaka, H. (2004). Contact-damage-resistant ceramic/single-wall carbon nanotubes and ceramic/graphite composites, *Nat. Mater.*, **3**, pp. 539–544.

501. Jiang, D. T., Thomson, K., Kuntz, J. D., Ager, J. W., and Mukherjee, A. K. (2007). Effect of sintering temperature on a single-wall carbon nanotube-toughened alumina-based nanocomposite, *Scripta. Mater.*, **56**, pp. 959–962.

502. Ahmad, K., and Pan, W. (2008). Hybrid nanocomposites: a new route towards tougher alumina ceramics, *Compos. Sci. Technol.*, **68**(6), pp. 1321–1327.

503. Balani, K., Zhang, T., Karakoti, A., Li, W. Z., Seal, S., Agarwal, A. (2008). *In situ* carbon nanotube reinforcements in plasma sprayed aluminum oxide nanocomposite coating, *Acta Mater.*, **56**(3), pp. 571–579.

504. Xia, Z. H., Lou, J., and Curtin, W. A. (2008). A multiscale experiment on the tribological behavior of aligned carbon nanotube/ceramic composites, *Scripta. Mater.*, **58**, pp. 223–226.

505. Estili, M., and Kawasaki, A. (2008). An approach to mass-producing individually alumina-decorated multi-walled carbon nanotubes with optimized and controlled compositions, *Scripta. Mater.*, **58**, pp. 906–909.

506. Sridhar, I., and Narayanan, K. R. (2009). Processing and characterization of MWCNT reinforced aluminum matrix composites, *J. Mater. Sci.*, **44**, pp. 1750–1756.

507. Flahaut, E., Peigney, A., Laurent, Ch., Marlière, Ch., Chastel, F., and Rousset, A. (2000). Carbon nanotube-metal-oxide nanocomposites: microstructure, electrical conductivity and mechanical properties, *Acta Mater.*, **48**(14), pp. 3803–3812.

508. Peigney, A., Laurent, Ch., Flahaut, E., and Rousset, A. (2000). Carbon nanotubes in novel ceramic matrix nanocomposites, *Ceram. Int.*, **26**(6), pp. 667–683.

509. Peigney, A., Laurent, Ch., Dumortier, O., and Rousset, A. (1998). Carbon nanotubes-Fe-alumina nanocomposites. Part I: influence of the Fe content on the synthesis of powders. *J. Eur. Ceram. Soc.*, **18**(14), pp. 1995–2004.

510. Peigney, A., Laurent, Ch., Dumortier, O., and Rousset, A. (1998). Carbon nanotubes-Fe-alumina nanocomposites. Part II: microstructure and mechanical properties of the hot-pressed composites, *J. Eur. Ceram. Soc.*, **18**(14), pp. 2005–2013.

511. Peigney, A., Laurent, Ch., and Rousset, A. (1997). Synthesis and characterization of alumina matrix nanocomposites containing carbon nanotubes, *Key Eng. Mater.*, **132–136**, pp. 743–746.

512. Bakshi, S. R., Lahiri, D., and Agarwal, A. (2010). Carbon nanotube reinforced metal matrix composites—a review, *Int. Mater. Rev.*, **55**, pp. 41–64.

513. George, R., Kashyap, K. T., Rahul, R., and Yamdagni, S. (2005). Strengthening in carbon nanotube/aluminium (CNT/Al) composites, *Scr. Mater.*, **53**, pp. 1159–1163.

514. Tokunaga, T., Kaneko, K., and Horita, Z. (2008). Production of aluminum-matrix carbon nanotube composite using high pressure torsion, *Mater. Sci. Eng.*, **A490**, pp. 300–304.

515. Kwon, H., Estili, M., Takagi, K., Miyazaki, T., and Kawasaki, A. (2009). Combination of hot extrusion and spark plasma sintering for producing carbon nanotube reinforced aluminum matrix composites, *Carbon*, **47**(4), pp. 570–577.

516. Dong, S. R., Tu, J. P., and Zhang, X. B. (2001). An investigation of the sliding wear behavior of Cu-matrix composite reinforced by carbon nanotubes, *Mater. Sci. Eng.*, **A313**, pp. 83–87.

517. Kim, K. T., Eckert, J., Menzel, S. B., Gemming, T., and Hong, S. H. (2008). Grain refinement assisted strengthening of carbon nanotube reinforced copper matrix nanocomposites, *Appl. Phys. Lett.*, **92**, pp. 121901_1–121901_3.

518. Li, H., Misra, A., Horita, Z., Koch, C. C., Mara, N. A., Dickerson, P. O., and Zhu, Y. (2009). Strong and ductile nanostructured Cu-carbon nanotube composite, *Appl. Phys. Lett.*, **95**, pp. 071907_1–071907_3.

519. Goyal, A., Wiegand, D. A., Owens, F. J., and Iqbal, Z. (2006). Enhanced yield strength in iron nanocomposite with *in situ* grown single-wall carbon nanotubes, *J. Mater. Res.*, **21**, pp. 522–528.

520. Jeong, Y. J., Cha, S. I., Kim, K. T., Lee, K. H., Mo, C. B., and Hong, S. H. (2007). Synergistic strengthening effect of ultrafine-grained metals reinforced with carbon nanotubes, *Small*, **3**, pp. 840–844.

521. Dong, L., Subramanian, A., and Nelson, B. J. (2007) Carbon nanotubes for nanorobotics, *Nano Today*, **2**, pp. 12–21.

522. Huang, X., and Zhang, S. (2010). Load-driven morphological evolution in covalently bridged multiwalled carbon nanotubes, *Appl. Phys. Lett.*, **96**, pp. 203106_1–203106_3.

523. Zhu, T., Li, J., Ogata, S., and Yip, S. (2009) Mechanics of ultra-strength materials, *MRS Bull.*, **34**, pp. 167–172.

Index